The Latent Order of Complexity

THE LATENT ORDER OF COMPLEXITY

JOSEPH P. ZBILUT
AND
ALESSANDRO GIULIANI

Nova Science Publishers, Inc.
New York

For permission to use material from this book please contact us:
Telephone 631-231-7269; Fax 631-231-8175
Web Site: http://www.novapublishers.com

LIBRARY OF CONGRESS CATALOGING-IN-PUBLICATION DATA

Available Upon Request

ISBN: 978-1-60456-026-8

Published by Nova Science Publishers, Inc. ≃ New York

DEDICATION

To my family—per ardua ad astra—Virgil
J.P.Z.

To my father from whom I learned not to take my ideas too seriously, and to my wife who still reminds me of this lesson.
A.G.

CONTENTS

FOREWORD

Since the beginning of the 20th century, “How Things Work” has appeared as title or theme of numerous popular books that try to dispel the mystery, to the uninitiated, of the forms and functions of the bounteous assortment of “gifts” to our well-being, comfort and entertainment, provided by science and engineering. But, today even tinkerers who long ago could tune their own automobile engines, by cleaning points, gapping plugs and replacing the condenser, are severely challenged when faced with fuel-injected, stratified- charge engines monitored by on-board “chips”. Similarly, in medicine, for some major operations (e.g., gall bladder removal, prostatectomy) robotic surgery already outperforms the classical “wide exposure, hemostasis, gentle handling of tissues” that were the routine demands put upon surgeons in training who had to rely on their own hands.

The scope of public ignorance concerning how things work inevitably grows explosively. It is unreasonable to expect widespread or detailed understanding even of the many major support systems that make urban life possible (clean water, electrical supply, groceries in markets at all seasons, trash and sewage disposal…). What we don’t understand seems “complex” to us, at least until, with study or practice, we may achieve an occasional “Ah – Ha!” moment when complexity suddenly reduces to simplicity, and part of our world view changes forever.

In this welcome and appealing book the authors, who have achieved stature in both experimental and theoretical sciences, address the grandest “how things work” issue of them all, viz., the methods and limitations of science itself. They do so in a conversational style accessible to any interested reader. Queries about how science works have provoked controversies from the beginning, still continuing, but even more animated are debates about the limitations of science and its methods. Responsible and informed people are even writing books and articles with the theme: “The End of Science!” In that setting, Zbilut and Giuliani unflinchingly guide the reader through brambles and thickets, emerging unscratched – an admirable accomplishment.

A key issue for reaching understanding of physical sciences regards the relationship between empirical evidence (facts, data) and mathematical abstractions (models, theories) that “fit” the facts. In these sciences a successful approach has been to cast the text of a theory into a formal (i.e., mathematical) structure, with constraints or boundary conditions to specify it for particular circumstances, and to see if the resulting *model on theory* is compatible with a *model of data.* If so, the theory lives on for another test some other day. But, for the biological sciences this approach has not been triumphant. The grandest biological theory (updated versions of Darwin's theory) views the past as chronicles rather than as dynamics (expressing laws of motion and change so dear to the physical sciences). Freeman Dyson has recently reminded us that the physical sciences dominated the activities and philosophy of science until the middle of the 20th century, but since then the biological sciences have overtaken the physical sciences in importance, by the practical measures of funding support, attraction of young talent, and rate of discoveries.

Given these competing currents and trends within science itself, an informed, meta-level view, such as this book, provides needed re-orientation as we try to comprehend fast-moving events and discoveries accompanied by (often false) claims. Particularly valuable is the critique of determinism in science. The distinguished economist Kenneth Arrow once referred to the success of Newton's profoundly deterministic mechanics as "an intellectual disaster" because it became the almost universal model for motion and change. Zbilut and Giuliani add detailed support, in more moderate language, and argue for a piece-wise determinism that "may provide a way which recognizes the essential "building block" nature of many processes, while at the same time recognizing that building blocks are not always easy to assemble." In doing so, they pacify the warriors in "reductionism versus holism" battles.

In science it is often the case that between a concept and its realization, falls a shadow – a shadow that veils a simplicity lurking in the appearances of complexity. This book identifies that umbra, and like a good photo-processor, pulls hidden details out of the darkness.

F. Eugene Yates, MD
Professor of Medicine, and Ralph and
Marjorie Crump Professor of Medical Engineering (emeritus), UCLA
Scientific Advisor, The John Douglas French Alzheimer's Foundation, Los Angeles, CA
Los Angeles, CA
July, 2007

PREFACE

After a certain high level of technical skill is achieved, science and art tend to coalesce in esthetics, plasticity, and form. The greatest scientists are always artists as well.

Most of the fundamental ideas of science are essentially simple, and may, as a rule, be expressed in a language comprehensible to everyone.

Albert Einstein

Science occupies a peculiar place in human activity. We think nothing of the fact that a significant portion of daily existence is made to some degree easier because of science. Mundane contingencies such as laundry, cooking, sanitation, travel have all been made easier because of it—to the degree that we do not make the connection with science. Yet, if new ideas such as stem cells emerge, the discourse quickly becomes charged and cautionary. Even among professional educators, the topic can be the basis for significant passion. The philosopher, theologian, sociologist—and scientist all have views on what it is and what it should be.

We would like to add this monograph to the discourse. It is not intended to be a strictly scholarly work—it is intended for those with some degree of education, who are either scientists themselves, graduate students, or are interested in science. Judging from the space or amount of time the media devote to the topic, this may be a considerable number. Indeed, the most mercantile of interests may benefit, if once considers the frequent articles found in the *Wall Street Journal*. Although the book is best approached as a whole, each chapter can be profitably read alone. The notes at the end of a chapter are more for the purpose of making ancillary comments, than citations.

We would like to thank all our friends and colleagues who have at one time or another, consciously or unconsciously helped to form our view. We do not want to list their names for fear we will forget someone inadvertently.

Finally, it should be noted that our tone is not "serious" in the traditional sense, but playful and sometimes irreverent, in the manner one might have a discussion with a friend over coffee, or a glass of wine. In this sense it has been a pleasure writing it, and we hope readers will derive the same sense.

Chapter 1

SIMPLICITY IS DIFFICULT

Pluritas non est ponenda sine necessitate. [Keep it simple!]

William of Occam (Figure 1)

Simplicity is the ultimate sophistication.

L. Da Vinci (Figure 2)

Figure 1. William of Occam.

Figure 2. Leonardo da Vinci.

THE MOTIVATION FOR THIS BOOK

The debate about the role and status of science in our society is a significant concern for our times. Most importantly the concern spans all strata of society, provoking opinions from professionals, politicians and the common man. The significance of this concern stems from the fact that science has invaded areas very far from its original base to include religion, philosophy and political beliefs. This exaggerated and unusual role played by science has generated equally exaggerated and unusual reactions ranging from rigid opposition to equally rigid adoration.

In our opinion this is dangerous for both science and society: For science the risk is in being mummified in the role of "perfect truth," thus losing any possibility of evolution (this means the death of science); whereas, for society the risk is in being constrained by a unique way of thought where there is an "optimal" single solution to each problem, and no room for diversity (this means the simultaneous death of progress and democracy).

For all these reasons, in our opinion, knowledge of the basic principles of scientific enquiry has become particularly crucial even for people not directly involved in science. Professional scientists are too often preoccupied by so many minor peculiarities of their own specializations that they forget the basic pillars of their work, and consequently they lose the possibility of having useful contacts outside their very small, largely auto referential, circles. Certainly, the demands of highly specialized topics require students to become quickly immersed in the details of objects of their study, leaving minimal amounts of time to study broader viewpoints, or even related areas of science. The search for funding and positions precludes any reasonable "free" time to examine related issues. Even so-called "broad-based" undergraduate and graduate curricula are sacrificed for pragmatic exigencies. University degrees have slowly moved away from the idea of "education," toward a reinvention of mediaeval guilds or unions, whose function is to be a gatekeeper for the increasing "professionalization" of a variety of traditional trades. To be sure, such a professionalization is not without merit, and is largely a societal decision; yet, perspective is quickly lost, and science becomes whatever is deemed to be contemporary science. The distinction between science and engineering becomes increasingly blurred, and the popular media more and more become the arbiters of scientific significance. Certainly, science becomes less and less a rational discipline, and more and more a provider of "miracles."

An analogy is the distinction between an artist and an artisan: both are remarkable for their abilities and are highly regarded. However, it may be said that the distinction is between someone who can present a perspective faithfully and one who can present a perspective in a totally new and remarkable way. The Renaissance was populated by many who practiced the arts and the sciences, but few practiced it as well as a Galileo (Figure 3) or a Michelangelo (Figure 4). The supposed rivalry between Mozart (Figure 5) and Salieri (Figure 6) hint at this distinction. An important point to be made is that both the artisan and artist, engineer and scientist are necessary in some proportional way—they inform each other. To emphasize or let one overtake the other risks the existence of both.

Figure 3. Michelangelo.

Figure 4. Galileo.

Figure 5. Wolfgang Mozart.

Figure 6. Antonio Salieri.

This is particularly dangerous in our times where the mainstream way of doing science is rapidly losing its efficacy (the most evident paradigm of this crisis being the paradoxical negative correlation between the apparent increase in biological knowledge and the number of new marketed drugs) in facing the problems coming from the real world.

The aim of this book is to give an infinitesimal (but greater than zero) contribution to the goal of a larger appreciation of what scientific work really is. We start with the tools scientists use in their everyday work; to give a convincing justification for the need of a drastic simplification of scientific methodologies and approaches with the elimination of unnecessary burdens science has accumulated in the last three centuries of existence. We are convinced that the basic sense of scientific work can be conveyed to the layman without entering into the specialized content of single fields (a practically impossible enterprise) but, on the contrary, concentrating on the methods of reasoning employed by scientists.

The widespread diffusion of low cost and very powerful computational tools gives us the practical possibility to start this simplification. The obstacles are of a different kind: they have an intermingled philosophical, historical and sociological nature, and are deeply rooted in the structure of the scientific community that, like any big and powerful institution, tends to become more and more rigid and conservative with age. Our aim is to present, with the aid of some paradigmatic case histories of simple science (we hope the sense we give to the word simple will be apparent in the subsequent pages) a reasonable path to the elimination of the incrustations deposited on the keel of the science. A necessary corollary to this basic aim is that this book will convey our idea of science: neither a philosophy nor a religion but a fascinating work of craftsmanship that in some rare cases becomes artistic.

SETTING THE RULES OF THE GAME OR MICHAEL JORDAN'S SIMPLICITY

Ask a mathematician what he means by "simple," and he will probably show you a very concise formula that is supposed to explain many things (or better, "a very large class of problems"). This is a dirty trick, as any of us who attended a university-level course in mathematics knows, there is an abyss separating the first two pages of any mathematical textbook from the rest: apparently very simple assumptions and concepts turn into a completely unnatural formalism almost impossible to understand. We can call this effect "the curse of absolute rigor": in trying to develop an absolutely self-consistent and objective way of describing things we crash with the finite character of our day-to-day experience and intuition. Any general, self-consistent and context-free formalization of anything flirts with stupidity and the absurd. There is a joke that explains this point: motorists are fond of making fun of highway police officers, and the story goes that two officers are in a car, the first officer asks the other "Is the flashing light activated?"—the second answers "Now yes, now no, now yes, now no, …". [1]

If we look just one centimetre below the surface of this joke we discover that the answer of the second officer is perfectly appropriate if we forget the "context information" that a flashing light is supposed to have an intermittent behavior. Now think of what terrible fatigue could ensue if one had to consider all the possible exceptions and "context dependencies" of a rule to generate a self-consistent and autonomous language. The dream of simplicity turns into a nightmare. This is the reason why "looking for generality" is not a good way for finding simplicity. In a more serious vein, consider the problem of the statement "… , with x vanishingly small, .."

For those schooled in mathematics or physics this phrase is common and has a significant history, but for the average person, the phrase may be strange. Indeed, "vanishingly" suggests something mysterious or even magical—certainly not objective, or "scientific." But in the mathematical-physical world it deals with the basis of calculus. The concept came to be very useful for describing mathematically a variety of physical systems in very nice concise ways. Yet, if some of the assumptions involved are carefully examined, a number of strange things occur when compared to our own sensible observations.

Take for example, using the vocabulary of a physicist, the case of a one-dimensional motion of a "particle" decelerated by a friction force with respect to some equilibrium. In other words, we are trying to describe an object which is moving toward some resting point by being slowed down by friction. In mathematical terms it might be described by $m\dot{v} = F(v)$, where *m* is the mass of the object and v is the velocity. (There are many ways to express this mathematically—here we are simplifying the expression for the sake of the discussion.) The little point on the v indicates its "derivative," i.e., its rate of variation in time or, more concisely the acceleration. The way the mathematics works is that a single solution should exist to the problem. One can "linearize" the force *F*, with respect to the equilibrium,

$$v{=}0:\ F \to -\alpha v \ \text{at}\ v = 0,\ \alpha = \left(\partial F / \partial v\right)_{v=0} > 0\,.$$

The solution for this equation, meeting the existence constraint, as the object approaches the resting point is

$$v = v_0 e^{-\frac{\alpha}{m}t} \to 0 \ \text{at}\ t \to \infty, v_0 = v(0)$$

It is not important to consider all aspects of the equation.[2] The significant part of the solution for our purposes is the expression, $t \to \infty$, which states that the resting point cannot be reached in a finite time! The usual explanation of such an effect is that, to the accuracy of our limited scale of observation, the particle "actually" approaches the equilibrium in finite time. But this begs the question—in fact the equation states it requires an "infinite" time to arrive at the resting place. The implication is that the equation is really a kind of joke, and that this little detail should be ignored, since our practical experience tells us otherwise.

This little difficulty recalls the famous paradoxes proposed by the Greek philosopher, Zeno of Elea in Southern Italy (Figure 7), and were introduced as demonstrations that motion was an illusion. As recounted by Aristotle (*Physics* VI, 9), the one most obviously related to the equation problem is trivially expressed as "to get there, you must first get halfway there." Thus, suppose you want to go from point A to point B. First you must go half the distance; then half the remainder; then half of that remainder; i.e., ½, ¼, 1/8, etc. You must in effect travel an "infinite" number of finite distances—an infinite time of travel!

Figure 7. Zeno of Elea.

The main problem for the equation is that it looks for a classical Newtonian formulation for a deterministic single solution. Without getting into the history of deterministic thought at this point, the solution is presented in the mathematical framework of Platonic idealism. The formulation of the problem is presented in the name of a "particle"—an amorphous unknown entity to provide for generalization. However, real life is not necessarily congruent with this view. Take as an example the act of hitting a nail with a hammer: each blow of the hammer roughly takes the same trajectory, but we know each trajectory is not exactly the same. Similarly, if we roll a ball down an incline several times, the trajectories are roughly the same, but not exactly the same. And, in fact, depending upon the angle of the incline the trajectories may be very different from each other—or very similar. The point is that as these real objects arrive at their resting point, they can do so by several different "solutions," and they do this in a finite time. Classical approaches do not handle these "messy" little details well because the classical solutions are based upon an idealized reality. There really is no "context," and consequently, the whole endeavor is made more complex by having to "explain" apparent paradoxes. (In fact, mathematics—and science as a whole, does recognize this problem by its need to "control" experimental variables.) In fact, there is a very simple mathematical solution to this problem: allow for multiple solutions. Those who are mathematically inclined may recognize this solution by a relaxation of the Lipschitz (Figure 8) conditions for uniqueness and existence. In general this "condition" means there is a unique solution to a differential equation.

Figure 8. Rudolf Lipschitz.

Figure 9. Michael Jordan.

Thus, it is important to underscore the fact that many absurdities and complexities in such scientific formulations are often a result of assumed idealizations which are not at all contextual. Certainly, mathematicians have implicitly known this for a long time as they often analyze "boundary conditions" for the formulations. Moreover, more recently, due to increasing knowledge about how our brains work, it is becoming realized that mathematics (and science) tend to be driven by our own habits of perception: a need to see simple patterns—even if none exist!

Convinced? Nevertheless mathematics in science is a blessing and not a curse, given the fact that we use it like a carpenter uses his hammer—not like an end in itself but to drive the nails. Scientific statements live a finite life in a finite world, and are born with very strict applicability ranges. To take them to the infinite invariably makes them silly and useless. On the contrary, infinity is where mathematics looks for its proofs. The simplicity of mathematics is fascinating but is something different from what we are looking for.

Ask a basketball player what he means by "simple," and he will show you the so called "fundamentals" of basketball which can be summarized by very basic linear actions: throw the ball to the basket, bounce the ball and make sure you throw it before you take a certain number of steps. That's all, this is simple, really simple, and can be explained in ten minutes to anyone. But what makes a Michael Jordan different from us? Well, two main things: a) the instinct of choosing from this very simple repertoire of actions the right thing to do in every possible situation; and b) performing each of these basic actions in the best possible way. This is the simplicity we are seeking: finding the minimal set of basic actions to perform, with each action being easy to understand. Then the play can start and the champion will be the one able to find the right thing to do when in front of any basically unpredictable and ever changing situation.

Obviously not everyone can become a Michael Jordan (Figure 9), but anyone can appreciate the difference between Michael Jordan and the guy next door when they are on a basketball court. The possibility to appreciate this difference in the game of science is the goal of this book.

THE GAME OF SCIENTIFIC FUNDAMENTALS

Science is a matter of rhetoric. Modern times consider rhetoric a bad word, but this was not the case in ancient Rome where rhetoric was simply one of the tools lawyers used to do their job. The tool of rhetoric (like any tool) is *per se* neutral, not good, not bad—it becomes bad when, let's say, we want to convince someone to spend their money in a business we know will certainly end in bankruptcy. It is good when we want to convince our two year old daughter to drink a very bitter tasting medicine we definitely know will improve her health. The illusion of completely self-consistent, self-explaining, self-evident language as the natural language of science, made some philosophers think that rhetoric has nothing to do with "objective science", but, in turn, this is very bad rhetoric. Negating the contextual dependence of scientific statements and achievements and their implicit finite nature has been a fraud in contradiction to the actual development of science: history is filled with pseudo-sciences like phrenology or pseudo-entities like phlogiston, whose existence was subscribed in good faith by the most eminent scientists of the time. Real science grows from errors. More generally, any thing with too many "self-" in its own definition sooner or later demonstrates its insufficiency as we all discovered in adolescence. This means that we must be conscious of the fact that we must actively convince other people of the soundness and relevance of our conclusions. There does not exist any scientific result that is true and convincing "by itself." If having stated rhetoric (i.e., the art of convincing someone else of the plausibility of your own statements) plays a preeminent role in science, let's see what are the basic "figures" of scientific rhetoric.

Perhaps the most basic figure of scientific rhetoric is the "Translation." Think of an epidemiologist trying to convince us of the effect exerted by a given chemical x in increasing the risk of lung cancer. He will subdivide a given population of individuals into two subsets: the subset T made of individuals exposed to the chemical x and the subset P made of individuals who were not exposed to the chemical during their life. Let's forget for a while the difficulty of ascertaining the correctness of the classification. The epidemiologist will use some "tricks" based on common sense to optimise this step. In any case, having made this dichotomy, we can immediately determine if any individual of the population is part of the T or P group.

Next we must perform another dichotomous classification on our population, namely to sort each individual into two groups "lung cancer suffering" (*LC*) and "others" (*O*). Now the game is to show how the *T/P* classification is mirrored by the *LC/O* classification, that is, to "translate" the knowledge of being T or P into a statement of being *LC* or *O*. Clearly the translation will be far from perfect. Not all T individuals will come out to be *LC*, but the epidemiologist will compare the joint distribution of individuals in the *T/P LC/O* categories with the distribution corresponding to the random assortment and will produce a statement about the "distance" of the observed distribution from what is expected by chance alone.

Chance is made to correspond to a perfect symmetry of assortment of individuals in the classes on the basis of the natural prevalence of each condition. Clearly the epidemiologist could have worked in the opposite way, i.e., collecting an equal number of *LC* and *O* individuals and then looking for their T or P condition. Like any translation it works in two directions. The convincing effect of such a procedure on the audience will be judged on the complete symmetry of the two subsets for all the possible causes different from exposure to the chemical x. Let's say the T category could have a different distribution of age with respect to P and thus the supposed effect of the chemical being an artefact of the effect of age. People of the T category could have different smoking habits and so on. Here the basic point to remember is the need to "express something by means of something else" and thus predict a disease by some life habit—the melting point of a molecule by its structural formula, the structure of a protein by its sequence, etc.

The second (and last) scientific rhetorical mode is "Measurement." Clearly, most people believe they understand measurement; after all, it is something we routinely do, such as determining the width of our waist, the distance on a map, or even figuring out a budget. But it is not that simple. What do we mean by a meter? Of course we are referring to some agreed upon standard—which also has a relation to a centimeter. These kinds of implicit relations had to be worked out, and probably developed as a result of social intercourse. We all recall historical examples of civilizations developing units of measure based upon, e.g., length of a person's foot or some other anatomical feature—a purported common reference point.

But later, especially as the notion of "scientific" observation developed, it became important not only to refer some object to some number, but also determine how well it did this; i.e., what kinds of errors could be introduced. More importantly could the measurements also be referred to some specific observable feature whose measurement could be related to other numerical measurements. The trivial example is: could one kilo of onions be equated to one kilo of turnips to produce two kilos of vegetables in a bag? And what kind of errors would I introduce by weighing the vegetables with a spring balance versus with a weight balance?

Ultimately, what we see is a gradual refinement of referential rules which relate a variety of naturally occurring objects to specific numbers. And still further, to manipulate those numbers to provide derived insight into the objects. This is perhaps the nexus of confusion between complexity and simplicity: the numbers are abstractions of perceived attributes.[3] Manipulation of the abstract world is convenient and efficient, but the abstract world is not the physical world. This is a simple and understandable statement, yet it has engendered numerous arguments from medieval philosophers to contemporary physics.

Let there be no misunderstanding. The simplicity we speak of is a simplicity difficult to achieve and it involves a clear appreciation of what is done—not simply the application of simple formulas. In the subsequent chapters of this book we will try to put these very general statements more explicit by commenting on different pieces of science. Consequently this book assumes a bit of an unusual "spiraling" trajectory (which could be disturbing for some readers to whom we ask for their forgiveness in advance) in which the same basic concepts will be recurrently presented under different forms and in different contexts.

REFERENCES

[1] Giuliani A, Zbilut JP (1998). The relevance of physical and mathematical modes of thought on complex systems behavior in biological systems. *Complexity* 3(5): 23-24.

[2] For a more detailed, mathematical treatment see Zak M, Zbilut JP, Meyers RE (1997). *From instability to intelligence: complexity and predictability in nonlinear dynamics.* (Lecture Notes in Physics: New Series m 49). Springer Verlag, Berlin Heidelberg New York.

[3] In his interesting book, *The Poincare Conjecture*, Donal O'Shea (New York: Walker Books, 2007) brings up the interesting hypothesis of Lucio Russo (*The Forgotten Revolution: How Science was Born in 300 BC and Why It Had to Be Reborn*, trans., Silvio Levy, New York: Springer, 2003) that the *Elements* were a theory of computation developed as an analogue to ruler and compass—like a slide-rule. Thus there is the suggestion that the ideal was not always the object of mathematics. Which then brings to mind the "non-Euclidean" geometries of the 19th century which herald complete idealization.

Chapter 2

SCIENTISTS ARE ARTISANS, NOT PRIESTS

Poets say science takes away from the beauty of the stars - mere globs of gas atoms. I too can see the stars on a desert night, and feel them. But do I see less or more? The vastness of the heavens stretches my imagination - stuck on this carousel my little eye can catch one - million - year - old light. A vast pattern - of which I am a part... What is the pattern, or the meaning, or the why? It does not do harm to the mystery to know a little about it. For far more marvelous is the truth than any artists of the past imagined it. Why do the poets of the present not speak of it? What men are poets who can speak of Jupiter if he were a man, but if he is an immense spinning sphere of methane and ammonia must be silent? "

Richard Feynman (Figure 1)

Figure 1. Feynman (center) with Oppenheimer (right).

A major motif pervading the history of mankind has been humanity's search for meaning. By this we mean a consistent effort to understand "reality," and how mankind fits into it. A corollary to this is humanity's efforts to control this reality. Obviously, religion has been an integral expression of this quest. From the so-called primitive to the most sophisticated of cultures, all have expressed some form of this pursuit. In the Western Tradition, with the rise of the Renaissance, a significant departure from traditionally exclusive religious approaches occurred. We all know this as the rise of the scientific method and its attempts to understand the perceived world. More or less coincident with this departure, the philosophical apparatus underpinning this movement also developed. The notion of reason became partialed out of the broad discourse involving reality. Theology was sequestered as a separate discipline. Without making a value judgement, it should be noted that idealism as a subject to be understood apart from the observable world was given formal definition by the likes of Descartes (Figure 2) and Kant (Figure 3). The practical result for our age is the recognition of "science" as the vehicle for progress due to its many successes. Because of the objectivity and impartiality of its methodology, it is seen as perhaps one of the most important aspects of modern life. Although most people have little practical knowledge of science, they are willing to admit its importance.

We would argue, however, that the "break" between science and theology has never been complete—at least in the popular mind (again here we make no value judgments). On many occasions we hear about "scientific truths," or its "miracles"; moreover, many scientists describe their work as "seeking truth." Much of this vocabulary may be a residual effect from the times when reality was solely interpreted by religion and its officials, priests. There tends to be a strong connotation of mystery embedded in looking at science as "searching for truth." This is not only false, but dangerous.

Figure 2. René Descartes.

Figure 3. Immanuel Kant.

During the 1970s, Ivan Illich (Figure 4) appeared briefly as a radical critic of contemporary institutions. His works were quite popular, but have since been largely ignored. Although many of his opinions seemed at times to be contradictory, he was basically a critic of contemporary idealism as embodied in modern institutions. In his *Medical Nemesis* he asserted that physicians had replaced priests as masters of death [1]. Although many have argued that his perspectives were too extreme and unrealistic, some have recognized that medicine, as perhaps the most pervasive impingement of modern science into everyday life, had taken over the absolute pronouncements of faith required of church members. In other words, medicine has become the temple of faith and correct living. And it should be remembered that today's medicine is really a joint effort of the entire scientific establishment. Not only do the physical sciences play roles with such tools as radioactivity, but obviously the chemist, and molecular biologist. And interestingly, when surveys indicate that popular interest in scientific education is not strong in many western settings, the media have increased their production of popular reports of science—especially medicine. The result is parallel to the behaviour of the medieval faithful who depended upon the Church for guidance in everyday matters. Today's "faithful" are the consumers of medical miracles packaged in media ads that promise a relentless "battle" against death. The "virtues," then, of this church are to "fight" bravely against disease [2].

Figure 4. Ivan Illich.

At the basis of the scientific endeavour is its evolutionary nature. This is to say that it is always refining itself, and that its interpretation of the observable world is never, in a sense, complete. The evidence for this is obvious: the history of science is replete with examples of theories that were at one time embraced and then discarded. This is beneficial. If a theory is to be believed, it must constantly evaluate its findings and determine if they are valid or not. The difficulty arises when findings are accepted without skepticism; i.e., if they are accepted as monolithic pronouncements. Often this occurs not at the hands of the generating scientists, but by their interpreters who may not even be professional scientists. Indeed, a case can be made that many of the worst problems of the last two centuries from colonialism to racism have at their basis inchoate "scientific" theories. The crimes against humanity justified by the application of scientific theories that appeared perfectly sound (and for this reason termed as scientific truth) are endless. Here it is sufficient to mention social Darwinism giving a "scientific" basis to colonial expansion and racism.

And contemporaneously, it may be that circumstances are worse. The conveyance of "truth" appears to have been idealized in media. Specifically, as an example, consider the story of a statistician who presented his work handwritten on several sheets of paper—his superior was not impressed and dismissed it. The statistician went back and typed it on computer paper, and this time the data was greeted warmly and deemed very significant. Certainly, now the computer paper is more often a spread sheet. But it echoes the idea that Marshall McLuhan suggested some years back that the "medium is the message." [3] Or the idea that if it is in a book, it must be true. And now the internet seems to be creeping into this category with newer generations.

Figure 5. Sir Isaac Newton.

Science is simply "searching for what works." The only scientific enterprise mainly guided by theory was the atomic bomb, in all the other occasions theory came after the practical thing. This is not to say that some bits of theory are active here and there, but certainly from the beginning of the scientific revolution, real science was a pragmatic process. Ptolemy's model of the universe is a good example—it was based on the theory that the earth was the center—and worked very well for centuries. From a pragmatic view, no one seriously challenged it since it could so well predict the motions of the then known universe. It was only questioned after the recognition that small errors were inherent in the model. [4]

Although Newton (Figure 5) is given the credit for formulating the laws of motion, in fact, as his statement that he "stood on the shoulder of giants" asserts, many scientists in Europe had already begun to suggest these laws as they derived them from a practical analysis of the world. Certainly Galileo is in this group. Newton was clever enough to bring these ideas into a coherent picture.

It is noteworthy that many ideas credited as "scientific truth" were shown to be completely false like flogiston, the age of earth as described by thermodynamics, the presence of electrons in the nucleus, the one gene-one protein dogma of molecular biology and so forth. Simply speaking, if something is termed as "scientific" it cannot in the same time be a "truth" because real science is intrinsically prone to an endless refinement and, more importantly, because science deals with measures that are programmatically partial and based on limited views of the world.

Although the "scientific revolution" is often described as an abandonment of metaphysical primacy in favour of empirical observables, this is not quite correct. Science deals with the measures of observables. And these measures must be common to different members of groups. Consequently, a physical object may be studied for its temperature, volume, mass, etc. These factors can then be manipulated to determine if there are any relationships among these. If these relationships are consistent, they may become the basis of a broader understanding of them so as to "predict" their behaviour. Thus the well known Ideal Gas Law describes the relationship of volume, pressure and temperature of gases relative to moles. But note that this does not, in fact, do away with abstractions.

This is perhaps most famously seen in Newton's force equals mass times acceleration: mass and acceleration can be measured, but force is an abstraction operationalized by the relationship between mass and acceleration. The same may be said of other concepts such as "work" or "energy." These concepts are abstractions operationalized by physical observables. Psychologists have sought to make the study of the mind "scientific" by operationalizing various concepts; i.e., by defining the concepts by behaviors.

Many scientists have noted the curious fact that these scientific "concepts," operationalized by mathematics bear a strong relationship to the Platonic philosophy of ideals. Indeed the noted statistical physicist, David Ruelle (Figure 6) [5], and mathematician David Mumford [6] have pointed to this and wondered about the "connection" with the human mind vis-à-vis the world of observables. To an extent, it may be argued that our perceptions are dependent to a degree on the categories of our minds, and that we operationalize them in a specific way. More and more studies have shown that the human mind tends to look for patterns or abstractions as a way to efficiently store our perceptions. Insofar as they work, they are useful. If they do not, we seek to refine them.

Nevertheless science, like every human activity, has rooted in its nature the "search for a sense" that is a form of "path to the truth." Let us try to define what is the sense of science, its specific path to the truth—in other words what are the "long lasting values" embedded in scientific activity. These elements allow us to recognize the role played by science in promoting and affirming human dignity.

Figure 6. David Ruelle.

Figure 7. Caravaggio's *Matthew's Vocation.*

The particular "truth" embedded in science is, in our opinion, the same kind of truth embedded in the work of artists and artisans. There is a beautiful collection of short stories by the American novelist Raymond Carver entitled "What We Talk About When We Talk About Love." In this beautiful title we can grasp the essential definition of what can be termed art: art is everything talking about "something else" with respect to its apparent meaning. This is the reason why real art has a much longer duration than its immediate motivations: when we look at Caravaggio's paintings we still derive deep sensations and resonances with our personal life even if we do not know anything about the problem of "Grace" in the Catholic Reformation (which was the main idea at the basis of Caravaggio's work) (Figure 7).

In the painting (Figure 7) displayed in the S.Luigi dei Francesi church in Rome, Jesus makes a personal call to the central figure, Matthew, and seems to ask "Is it me?". Grace is symbolized by the light coming from Jesus' side (not a terrestrial light—note that the light comes orthogonally with respect to the window). On the other hand, Grace is personal—Matthew hears the call whereas other people, like the young man at the end of the table, continue to count money without seeing anything. Even if we ignore all these particulars we still can grasp the basic meaning of the scene and we are captured by the immediacy of the painting.

The reference to "something else" with respect to the primary and self-evident use of a specific object is common to any artistic product. For example, when we look at an old wooden sailing boat, or at the architecture of an ancient loom, we can grasp the sense and logic of that object, even if these tools have a performance inferior to most modern ones. The sense comes from the particular structure/function relationship of the design—from the cooperation of all the parts toward the general goal—from the little singularities that make each specific object unique and speak to us about many “other things” with respect to the primary use of the observed tool. Mainly, these objects tell us of human work, ingenuity, fantasy, passion and efforts embedded in that particular piece of material.

Figure 8. Kandinsky's *Autumn in Bavaria*.

Mass production, eliminating singularities and constraining the object to a “single homogenized optimization” manufactures a drastic “reduction” of the senses by means of a massive de-complexification of the object. This de-complexification does not correspond to a simplification in the sense we are proposing in this book, but to its exact opposite. While simplification derives from the recognition of the multiplicity of environmental (physicists love to refer to environment as “boundary conditions”) demands and to the subsequent needs to avoid too many unjustified assumptions, de-complexification comes from the quest for a single “essential” characteristic of the studied system to be substituted for the real thing, and optimised without considering all the other intervening interactions.

Since the link between the Industrial Revolution and the “birth of ugliness” has already been investigated by other authors (Figure 8) [7], here it is sufficient to point the attention of the reader toward the beauty of nineteenth century Italian popular architecture (some world-wide famous examples being the Amalfi coast, the so called “trulli” in Puglia (Figure 9) or the “masi” architecture in the Dolomiti region) as compared to modern popular forms. Generally, contemporary builders are much more educated than earlier ones, but they share a mass “single optimum” view of their work that completely eliminates the beauty (and the sense) of their products: they destroy the landscape because they completely miss the link with specific boundary conditions. While their ancestors were connected to their localities (i.e., boundary conditions), by virtue of the regional character of their skills, modern architectures seem to have lost the sense of the “local.” Until now, science was the home of a very fine artistic culture: the specific skills with which any single scientist solves his/her local problems are a sort of signature, the style of reasoning, of applying methodology, of explaining the logical lines of the work. Thus we learn much from pieces of science which in many cases are already surpassed by new acquisitions and theories. A striking example is the so-called Pearson's product moment correlation coefficient, introduced to prove crazy eugenics theories between the 19th and 20th centuries. Nevertheless, not only is the coefficient still a great source of inspiration in our day-to-day statistical work, but the coefficient itself continues to be the most common metric for dealing with contemporary data mining problems in computational biology, as well as in other fields of science. The "added value" that keeps Pearson's r correlation coefficient alive and well despite the fact that the theory it was built for is dead, derives from the fact it talks of "something else" and incorporates the honest effort of an artisan to reach a goal.

Figure 9. Trulli in Puglia.

Figure 10. *Tower of Babel* by Brueghel.

This is no longer true for mass "science" in which myriads of blue-collar scientists work for a tiny piece of a greater picture the full meaning of which they do not understand. They simply apply previously established "standard methods" essentially adding another brick to a Tower of Babel (Figure 10) whose design came from some other place. But Babel towers have the dangerous tendency to fall. This was the case of the completion of the human DNA sequence (Figure 11). The "big idea" behind human genome sequencing was that the DNA sequence constitutes the repository of all relevant information about organisms, while discarding all the rest (zoology, botany, physiology, biochemistry) as pure accidents: the organism can be reduced to its DNA sequence where the ultimate causes of diseases, individual variability, and so forth, could be found. The one gene—one protein central dogma was the key of this reductionist view with the addition of the corollary that there is one protein–one phenotype. An army of "blue collar" scientists devoted years of work for this enterprise, and, at the end, the proteins were discovered to largely outnumber the genes due to completely unforeseen RNA alternative splicing processes. The same DNA sequence was found to give rise to different RNAs for the action of different RNA polymerases. These discoveries (and many others such as differential RNA degradation rates, junk DNA and so forth) completely demolished the basic assumptions of the enterprise from the "genetic material" side. On the "protein" side, the discovery of the rich dynamics of proteins with abrupt changes of function (phenotype) of the same sequence due to apparently slight changes of chemical microenvironment (shifts in pH, ionic strength, molecular crowding) and the correlated misfolding diseases (Parkinson, Alzheimer, scrapie) completely outside the realm of DNA control, destroyed the de-complexification efforts on the phenotype side. Using a metaphor, the genome passed from a classical music score faithfully executed by the players strictly following the music score to one played by jazz musicians as the basis for their improvisations: the DNA king was in the dust.

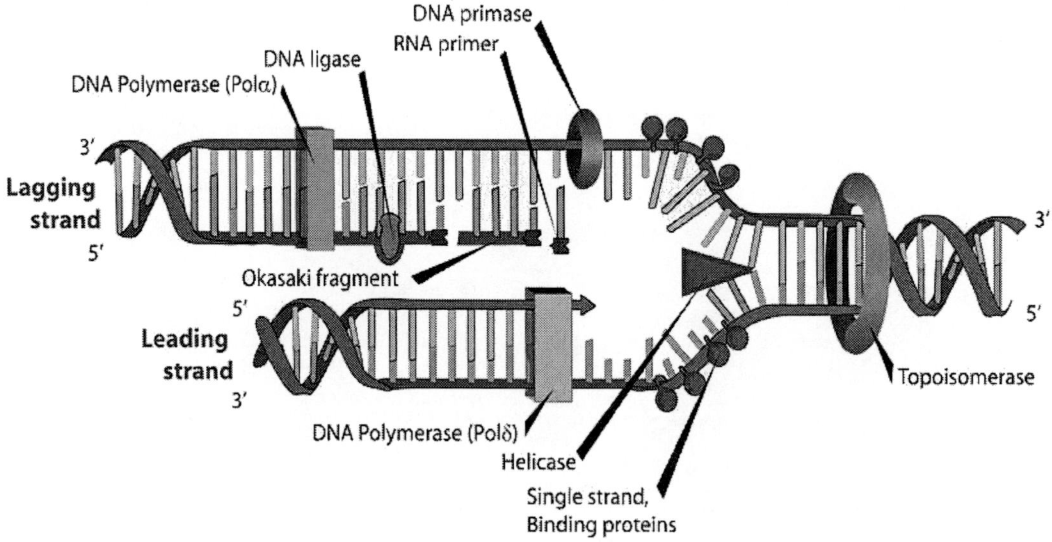

Figure 11. DNA Replication.

At the end of the day the Tower of Babel fell, destroying the dreams (or better, nightmares) of "total control" by genetic engineering that motivated the human genome adventure. This was a very healthy lesson in humility. Clearly some still try to maintain the dream by looking at other more sophisticated forms of control, but the scientific community is arriving at the conclusion that any arbitrary effort of reduction of the complexity of living to only one "Holy Grail" following the style of theoretical physics is bound to fail.

When a large building collapses, there is a lot of debris lying around from single bricks to still more or less solid walls. Something like this happened in Rome after the Gothic wars at the end of 6th century AD, when a city of one million people collapsed to a two or three thousand person village. From archaeologists' work we know very well what happened: people adapted structures built on a city scale (big temples, theatres, docks) to a small village scale (small houses, churches) by a new use of old pieces. This recycling involved a completely new mind and a free spirit in finding new functions for the different elements—a classical artistic task. The debris of the human genome project is now in the World Wide Web in the form of protein sequences, differential gene expression data, metabolic maps, etc. This is a perfect territory for artisan scientists looking for new uses of this material, i.e., new views of this material: for this task they do not need a lot of money or sophisticated instruments, they only need fantasy, statistical skills and creative spirit, which is the realm of simple science.

A subtle point underlying these observations is the fact that there seems to be a tendency to label "large" undertakings of processes as "complex," whereas, in fact, these processes are basically "simple." For example, DNA, although it contains tremendous amounts of material for coding, is essentially built up of only four base pairs. Apparently, its complexity is a function of its size. A massive building is still built of individual bricks (or steel beams, or window plates, etc.). And in this sense, perhaps the whole notion of "complexity" is really a result of human perception; i.e., the bigger, or more massive a thing is, the more likely it will be considered complex. This is something a nine year old child can easily understand. Children see the possibilities, whereas our adult minds, riddled with details, see only difficulty.

The Nobel laureate, and artificial intelligence pioneer, Herbert A. Simon (Figure 12), made some interesting observations in this respect in his *Administrative Behavior* [8]. He pointed out that human decision making is dependent on boundaries. Contrary to the belief that people gather all available facts before making decisions, human cognition was much more pragmatic: to try and assemble all the facts and then make a rational decision is almost impossible for the human mind—it takes too much time or may be confusing. Instead, the human mind is willing to take a less "optimal" result based on a pragmatic assessment of rational self interest. Implicit in this is the idea that if some problem is too big or "complex," a scaled down version of the model is created. Implicit also in this is the idea that if there were sufficient time and resources, an optimal solution would be feasible.

Figure 12. Herbert Simon.

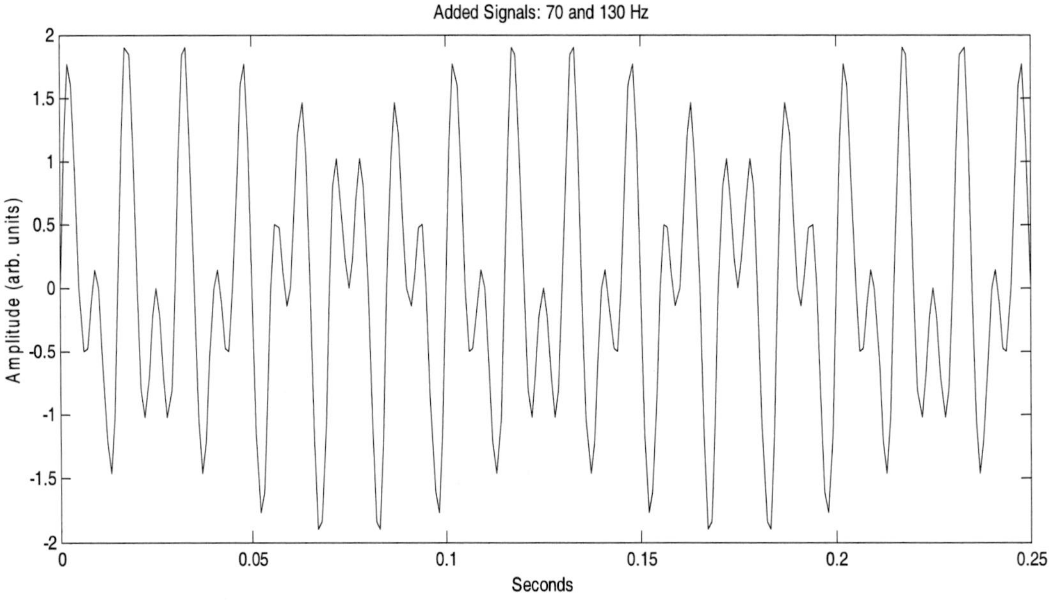

Figure 13.

The implications for simplicity vs. complexity would seem to be obvious. It's not that something is complex; it's that there is a lot of it. Complexity seems to be a surrogate for processes, ideas, structures, etc. that are massive and consequently are built up of many interconnecting smaller pieces. But strictly speaking, if one looks at the smaller pieces, they are often quite "simple." This is evident in the construction of a large building, human neuronal connections, cellular metabolism, etc.—it's a matter of scale. If one looks at the brain, it becomes immediately apparent that from a purely structural perspective, it is very simple: neurons, albeit of several kinds compose the brain. The difficulty is that there are millions of them, connected in a variety of ways. And moreover, they are signalling to each other very often simultaneously. The signalling itself is simple—if one knows the "language." Consider two simple sine wave signals—one at 70 Hz and one at 130 Hz being transmitted simultaneously (Figure 13).

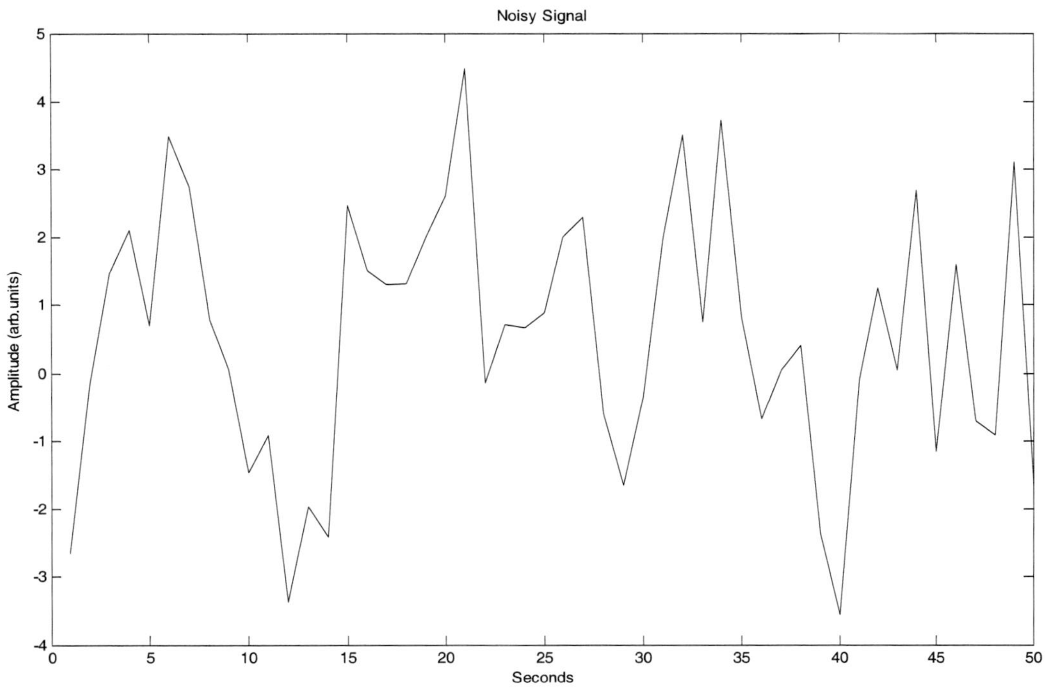

Figure 14.

It's fairly obvious that the signal has more than one signal present. And with some patience, one could figure out what those signals are. Unfortunately, real life is not that clean and simple. As any scientist can assert, signals are usually accompanied by "noise." Noise itself is difficult to define, but practically, it is all that material which is not really connected to the signal, but is unavoidable. It can be "misbehaving" cells discharging; stray signals from other areas, etc. The point is that the noise can make understanding the signal more difficult. If we add some random noise to the above signal, it might look something like Figure 14.

Clearly, the signal becomes much more difficult to understand, even though there are simply two separate frequencies involved. Modern mathematics has come to the rescue with the Fourier transform, which can decipher the signal. But even then, one has to recall that many signals may be superimposed as is found in a typical EEG (electroencephalogram) which records globally all the signals being generated. At this point, it becomes a very large problem. Furthermore, there may be "sub-nets," smaller groups of neurons forming a specific association. Thus the signalling between two neurons becomes something else, when their "discourse" is enlarged.

And because it's a matter of scale, it is for this reason, a large city such as Rome can be rebuilt of smaller pieces, and that the genome and proteins can be rebuilt from its individual pieces. It must be emphasized; however, this is not a reductionism to see how it works: the "size-complexity" formalism is tied to the interconnections. Knowledge of the interconnections is necessary to understand the "massive" object. Rome was "rebuilt," but not in the same way it was originally constructed.

Thus it can be said that there is a latent simplicity in the complexity of large "things." And, moreover, this complexity is probably more a reflection of the "boundary conditions" of our human brains, than the "things" themselves. This point of the "boundary conditions" is strictly linked to the artistic character of science: ideally a pure "rational" and "abstract" science in which laws are the most important elements does not need any specific skill by the user. It tends to become a completely automatic affair that can be executed simply following automatic recipes. The artisan specific skill is the ability to adapt preformed schemata to an ever changing reality that is never fully predictable and to the limitations of our possibilities. This is the specific ability of the scientist too.

REFERENCES

[1] (New York: Random House, 1976). In the book which was considered radical in its time, Illich severely criticized medicine for its lack of scientific basis. In some ways, the current drive toward 'evidence-based medicine" can be traced to his critique. He was diagnosed with cancer in the 1980s but refused all forms of therapy and died in 2002.

[2] Illich suggested that a result of medical consumerism would be a paradoxical increase in the usage of the medical apparatus. At the same time even health professionals have become to some degree alarmed at the all-pervasive nature medicine has become. Dr. F.T. Fitzgerald in her "Sounding Board" article in the *New England Journal of Medicine* ("The Tyranny of Medicine," 331: 196-198, 1994) points to a subtle change in the definition of "health" following WWII, in which the "norm" of health became the "ideal" of health. Thus sickness and accidents become faults. "So we select certain forms of self-abuse as deserving of opprobrium, and some of us have even seen colleagues refuse to care for certain self-abusers (such as fat people, smokers, drinkers, intravenous drug abusers, or those afflicted with HIV infection), since these victims brought it on themselves by not behaving as we told them to. In this way we can blame patients for their illnesses and deny them both our compassion and our services... The thrust toward the rationing of health care is making the association between vice and disease a public policy." She maintains that "... we will establish a tyranny of health in which those who are unwell are assumed to have misbehaved (with certain areas of misbehavior forgiven and others condemned)." She goes on to paraphrase Illich: "that classifying all the troubles of humanity as medical problems is actually antithetical to true health, in that it limits the ability of people to learn to cope with pain, sickness, and death as integral parts of life. Health, he maintained, is not freedom from the inevitability of death, disease, unhappiness, and stress, but rather the ability to cope with them in a competent way. If this is true, then the more medicine and society direct individual behavior, the less autonomous, and therefore the less healthy, the individual may become."

Although this debate continues, at the same time, there is a strange societal undercurrent which appears to distrust science in some "rational" way. Paul R. Gross and Norman Levitt (*Higher Superstition*, Baltimore and New York: Johns Hopkins University Press, 1994) document what they perceive as an antiscience cultural attitude often emanating from the humanities.

[3] *The Medium is the Massage: An Inventory of Effects* with Quentin Fiore, produced by Jerome Agel; 1st Ed. (New York: Random House, 1967).

[4] Howard Margolis documents the many folklores surrounding Ptolemy and Copernicus during the Renaissance in his *It Started with Copernicus* (New York: McGraw-Hill, 2002).

[5] Mathematical Platonism Reconsidered. *Nieuw Archief voor Wiskunde* 5, 30-33 (2000).

[6] The Dawning of the Age of Stochasticity. In: *Mathematics: Frontiers and Perspectives*. Ed. V. Arnold, V. Atiyah, P. Lax, B. Mazur. AMS, 2000.

[7] Probably the most clear and prophetic essay on this theme was written by a great artist in 1910: Wassily Kandinsky "Concerning the Spiritual in Art." The author clearly identified the risks embodied in a pure immediate and materialistic appreciation of the work of artists without any relation to both a social and educational purposes. The modern theory of "Art pour l' Art" was the gravestone of an Art really appreciated by large masses and influencing people at large. Avantgarde architects of the XVII century (e.g. Francesco Borromini) had their ideas percolating on day-to-day architectures and directly influencing the aspect of houses of the poor--nowadays these aspects have nothing to do with O'Gehary or Calatrava "cathedrals in the desert."

[8] *Administrative Behavior: A Study of Decision-Making Processes in Administrative Organizations*, 4th ed. The Free Press (1947)

Chapter 3

COMPUTERS MAKE SCIENCE EASY

Science is what we understand well enough to explain to a computer. Art is everything else we do.

Donald Knuth

Science is the belief in the ignorance of experts.

Richard Feynman

The last twenty-five years have witnessed a minor revolution in teaching mathematical skills. Gradually, pen-and-paper approaches have given way to increased use of computers in instructing students about such topics as geometry and calculus. And in some sense this mirrors our culture's gradual change from an analogue to a digital world. In recognizing the benefits of computer-aided mathematical instruction, the "continuous" world of calculus is reformulated as discrete mathematics. At the same time, the tedium is relegated to what the computer does best—perform computation; while, the student's new freed time is allowed for the appreciation of concepts and theories behind the computations. Certainly, there still is a debate as to whether this change is beneficial, but nonetheless it is pragmatic. As is often pointed out, the computer is the best tool for tedious, exact, repetitious tasks. And in this sense, this is nothing new historically. Historians of computation point out that over the centuries there have been many attempts to "mechanize" computation. Logarithms and the slide-rule (Figure 1) are among the venerable antecedents.

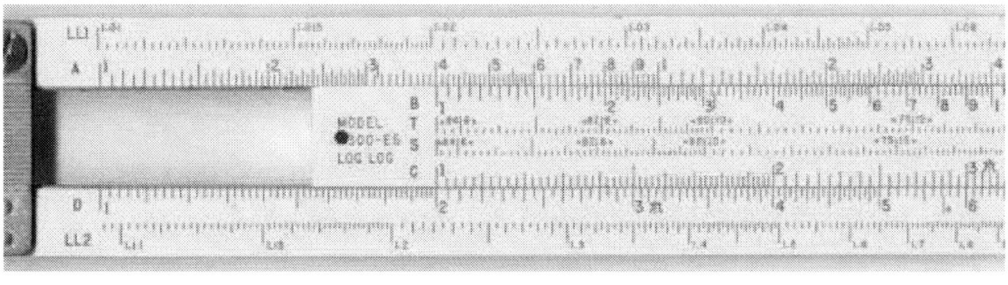

Figure 1. Slide-rule.

There is still another subtle benefit: if you do not need to make computations by pencil and paper, you do not need to learn the intricate formulas of calculus. It is sufficient that you learn the "general meaning" of mathematical operations (which often can be described by words and metaphorical representations) for using them. This implies you can trace back any scientific problem, in whatever field, to its "essentials" by a discussion with a "specialist," and then go to the computer and work. This paradigm happens over and over in scientific circles: "interdisciplinary" work becomes a matter of specialists relating to computational specialists from other fields. "Computational science" has grown more important as problems have become larger and larger, or when problems become relatively impossible to evaluate in the real world. Traditional boundaries of scientific disciplines disappear when physics, chemistry, and biology need to meet together to solve problems.

If you are a computational scientist, this strange hybrid living in the new niche made available by the possibility of having computational power at very low cost, means the "specialist" can be considered your client, in a way very similar to the patrons of Art of the Renaissance. If the client likes your quantitative representation of his results, he will be happy and have the impression of a real advancement in knowledge. This impression derives from the easiness of translating the mathematical objects you produced into his conceptual points in a way not basically different from the appreciation of a painting.

It is important to stress that this way of doing science is not simply confined to the direction of the modeller = mathematician / physicist → specialist = biologist / chemist, but can be equally applied in the reverse direction for the facilitating effect of the computer. The role of the modeller and the specialist (the painter and the sponsor) then becomes free of any academic specialization and simply dependent on the personal inclinations of the involved people. The subsequent example can help to clarify this point.

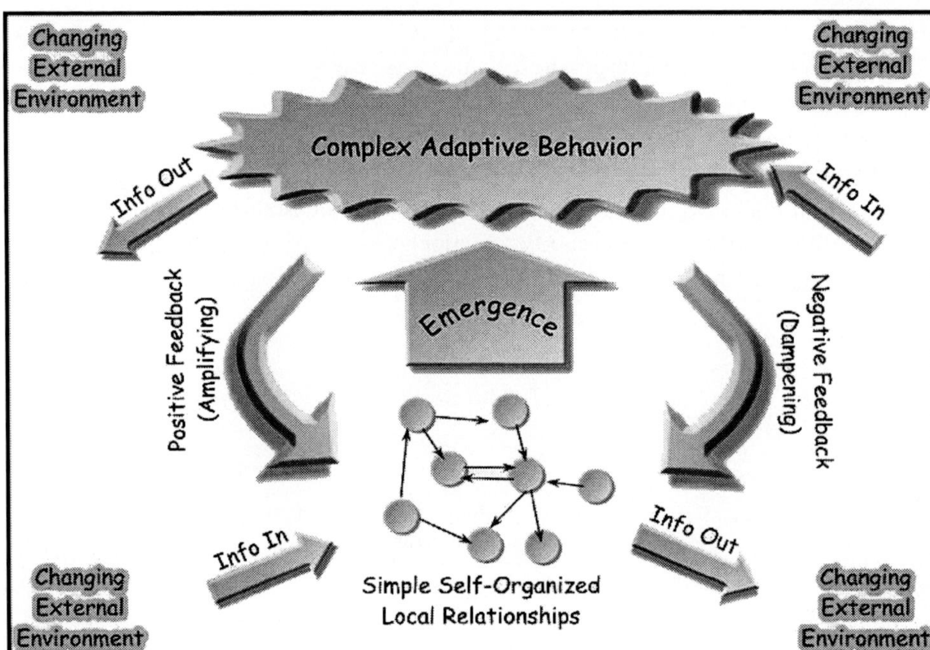

Figure 2. "Complex" System.

Over the last two decades, a significant debate has developed regarding the definition and use of the concept of complexity, inspired in some degree by interest in so-called deterministic chaos (a theoretical construct based on the observation of seemingly unpredictable and wild behaviors arising from very simple mathematical equations). A direct link between Chaos Theory and the real world was suggested by the analysis of time-series data in situations ranging from heart rate and brain activity to financial markets (Figure 2). A flourishing literature arose between the mid-1980's and the mid-1990's representing a sort of last attempt of strictly deterministic thought: the possibility to define the wild and unpredictable behaviour of a complex system. Confidence in these assertions diminished when questions arose regarding the applicability of the stringent assumptions of chaos detection methods to real world data.

Notwithstanding the unavoidable and rapid decline of Chaos Theory, some of the problems developed from this paradigm are still alive and worthy of attention. First of all there is the problem of measuring the "complexity" of a given entity (e.g., a time series, a network, a biological pathway or an organism). Roughly speaking, mathematically and physically oriented scientists, from their respective vantage points gave two kinds of answers to the problem of evaluating a system's complexity. Mathematicians developed the paradigm of algorithmic complexity; i.e., from the viewpoint of objects which somehow could be included in a metric defining complexity. This view looked upon the objects in varying terms in contradistinction to randomness. On the other hand, physicists defined the degree of complexity of a system starting from first principles, i.e., almost from an intuitive point. Thus if mathematicians were interested in randomness, physicists were interested in how the randomness was organized. In their basic characterizations, mathematicians were thus interested in the number of lines in a computer program as a system description; whereas physicists were interested in the dimensions of the system. Or alternatively, rules versus equations. Mathematical and physical approaches converged in the definition of complexity as the degree of unpredictability of a system. [1]

Both mathematical and physical recipes use the idealistic approach typical of hard sciences: a concept can be defined independently of the actual objects to which it is applied, so we can safely describe what velocity is [the derivative (rate of change) of the space with respect to time] even if we have no peculiar object moving. This kind of reasoning is very powerful on many occasions, allowing for very rigorous and general definitions. Clearly, this approach is untenable in biological sciences. For example, think of giving a rigorous definition of what a lymphocyte is even in the absence of white blood cells! Biological entities can only be defined in a bottom-up way trying to find relevant similarities (correlations) among existing individual entities. From Linnaeus onward, biology works generating operationally useful categories from the observation of similarities and differences: in this way we define organs, tissues, animal and plant species, chromosomes and so forth. And now this approach is experiencing greater refinement by the comparisons of protein and DNA sequences, metabolic network wirings, gene expression patterns and so forth.

Figure 3.

Physiologists analyze biological systems using reduced, simplified descriptions in much the same sense that an engineer uses a simplified description of an amplifier.[2] An engineer is often not interested in what is inside the box that produces gain (Figure 3). The engineer studies the properties of the amplification, but a complete description of the structure of the amplifier does not help much. A reduced description of the input/output relation of the amplifier is much more useful. In a sense, the goal is the "simple" understanding of the system. The internal structure of the amplifier is important, of course, when the engineer wants to improve or change the amplifier, but not so important when the goal is to use the amplifier or connect it to other devices to make a system.

An engineer presented with the idea that an unknown black box is an amplifier is like a biologist confronting an unknown biological system. Some structural knowledge is necessary. An engineer would have a terrible time if he were not told which connections were power supplies, which were inputs, and which outputs. But the last thing the engineer would want to know is the complete wiring diagram, let alone the locations of all molecules or atoms in its resistors, capacitors and transistors. Successful investigation requires some knowledge of structure; but it requires a number of measurements of inputs and outputs, under many conditions. Successful investigation also requires a good quantitative model of the system.

Physiologists have successfully analyzed a large range of biological systems using this approach. The kidney filters blood to make urine; the lungs transport oxygen from air to blood; muscles contract; sodium channels produce action potentials; etc. Each device description in physiology—from organ, to tissue, to cell, to organelle, to protein molecule—is associated with a model, just as a device description in engineering is followed by an approximate device model for its function, for example, its input/output relation.

On the contrary, a physico-mathematical top-down "concept first" approach regarding the degree-of-complexity of a model has sought to generate a set of "complexity measures" that could in principle be applied to real objects; namely, to experimentally observed time series constituted by arrays of values representing the time evolution of the most diverse systems (heartbeat, electroencephalographic traces, meteorological data, DNA and protein sequences). The problem is that, when applied to real systems, some of the basic assumptions of the theoretical definitions are violated: the real time series are too short, the stationarity constraints cannot be fulfilled, the series are not numerical, etc. Thus none of the proposed methods can be applied in a rigorous way to real data (again the need for artistic skill).

When some years ago [3] we approached this issue, we decided to turn the problem upside-down: instead of looking for a complexity definition that was both rigorous and applicable to real systems, we accepted the intrinsic fallacies of all the proposed methods as unavoidable. At the same time, however, we assumed that many of these non perfect methods captured a piece of truth—not the whole truth obviously, probably only a tiny piece, but different from zero.

The second assumption was that the different methods departed from "perfection" in different directions, i.e., they make different errors in defining the degree of complexity, but they are truthful in the same way. This is a very crucial philosophical point and has to do with how science has to deal with truth.

Whenever one speaks about "truth," one is always faced with the voluminous philosophical history of what "truth" is. Scientists, on the other hand, tend to have little patience for semantic (and perhaps ontological) disputations. They tend to be more direct: truth is what objective evidence suggests. Objectivity, of course, tends to be debated as well, but the scientific community has long ago agreed that "objectivity" has to be tested via the need for replication and verifiability. This is to say that there is a recognition that individual objectivity must be checked by communal objectivity. Certainly, this still does not guarantee that perspectives cannot change, but it pragmatically checks to make sure that an individual scientist's objectivity is not "mistaken." Witness the elaborate attempts to verify the theories of such luminaries as Einstein—no scientist is immune from verifiability. Moreover, here we are speaking of the development of a possibly unbiased method to measure the relative degree of complexity of a given series of numbers, thus truth is intended in the "soft meaning" we introduced in earlier chapters as "something that works for a given goal." In this case, the goal is a satisfactory ranking of a set of time series from the most predictable to the most unpredictable; i.e., from the most simple to the most complex.

If proposed methods can be imperfect in a myriad of ways, but can be right in only one way, then the logical consequence is that the best measure of the degree of complexity of a system is the portion common to all the (singularly imperfect) already existing methods. Clearly, this approach assumes that imperfect is not a synonym for wrong: imperfection lives in the continuum, and you can have different levels (and qualities) of imperfection. Instead, in a logical truth table, to be wrong or right is a yes/no concept. The existing methods are imperfect, so it is clearly possible, starting from a mixture of methods, to distil the good from the bad in a way similar to the production of whiskey.

In order to make the common (right and useful) portion of the different existing methods reveal itself, we must provide a good training set of time series to apply the different methods. The training set should be made of series normally encountered in real experimental settings: they must be not too long and not too short, going from very regular to very irregular (Figure 4).

Another important criterion is the insertion in the training set of some "mathematical series" generated by the application of a generating rule decided by us so as to know *a priori* the relative degree of complexity of these standard series. These series will be used for testing the relevance of our results. [3] In our case this task was accomplished by inserting in the training set two groups of standard signals:

1. ran (i) group: these signals come from originally random series made of 1000 values. These values are then allocated in ordered patches of varying length from short to long (i.e. ran1= order the first ten values from the lowest to highest, then the second 10 and so forth; ran 3 = order the first 50 values from the lowest to highest, then the second 50…, ran5 = order the first 200 values..).

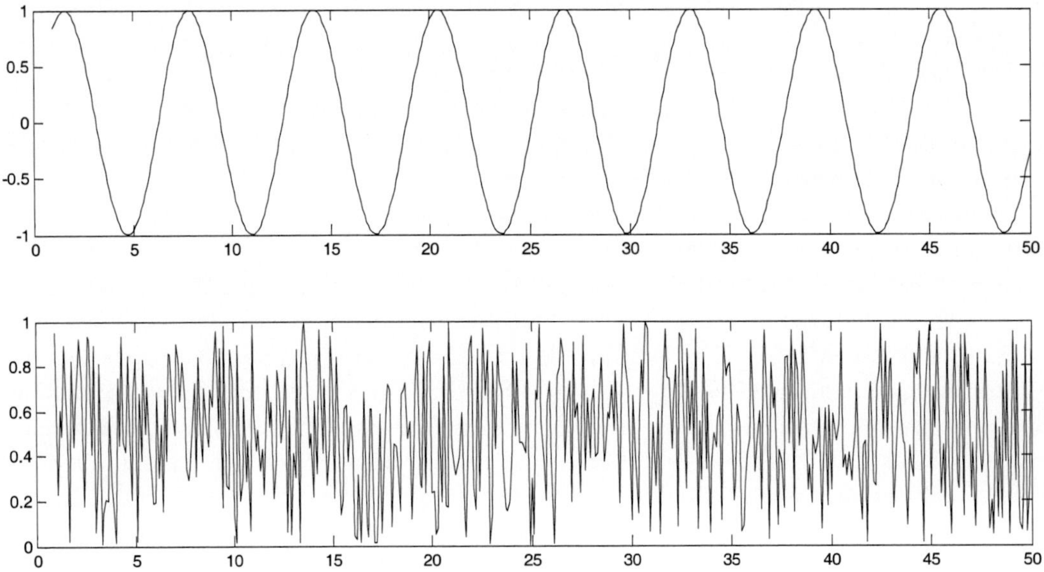

Figure 4. A very ordered sine wave (top), and random points (bottom).

2. snr (i) group: these signals come from the addition to an initially absolutely regular series made by the juxtaposition of a patch of 50 values equal to 10 to another 50 equal to zero values long patch and so forth until the point number 1000 of a random noise corresponding to a casual extraction from a Gaussian distribution. At increasing noise values going from snr1 to snr12 the series gradually loses its regular aspect to become very irregular. Figure 5 reports some of these standard series.

In analogy with the use of the term "standard" in empirical measurements (let's think of clinical laboratory analytical instruments) these signals allow us to judge the relevance of the complexity measures. A good complexity measure is one that reflects the ordering from the most regular to the most irregular we know already to be present in our signals.

Following the above criteria we collected a set made of approximately 200 series with the most diverse origins: federal funds, radioactive decays, tree rings width, rainfall records, DNA and protein sequences mixed with standard home-made series like the ones described above.

All these series were scored by means of 13 different theoretical complexity measures, coming from the application of different paradigms of complexity estimation. The data set having as statistical units the 200 series and as variables the 13 scores, was submitted to a classical multidimensional statistics method: principal component analysis (PCA). This is a time honored (it has been around, under different names since 1873) procedure with applications ranging from theoretical physics to psychology and market investigations. Without going too deeply into the explanation of the method, here it is sufficient to say that its goal is to extract, from a set of measures (variables) their correlated (common) portion so as to generate one (or more) synthetic scores summarizing the shared information present in the studied measures. More precisely the first principal component is the cumulative score that best represents the information shared by all the considered measures.

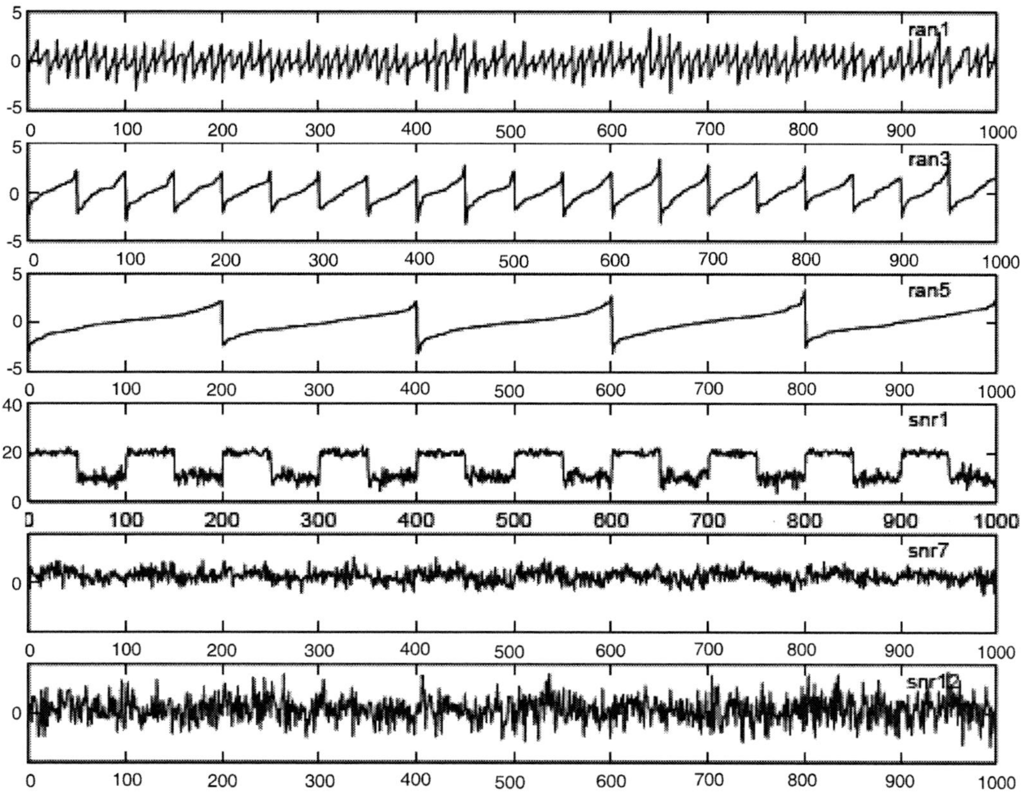

Figure 5.

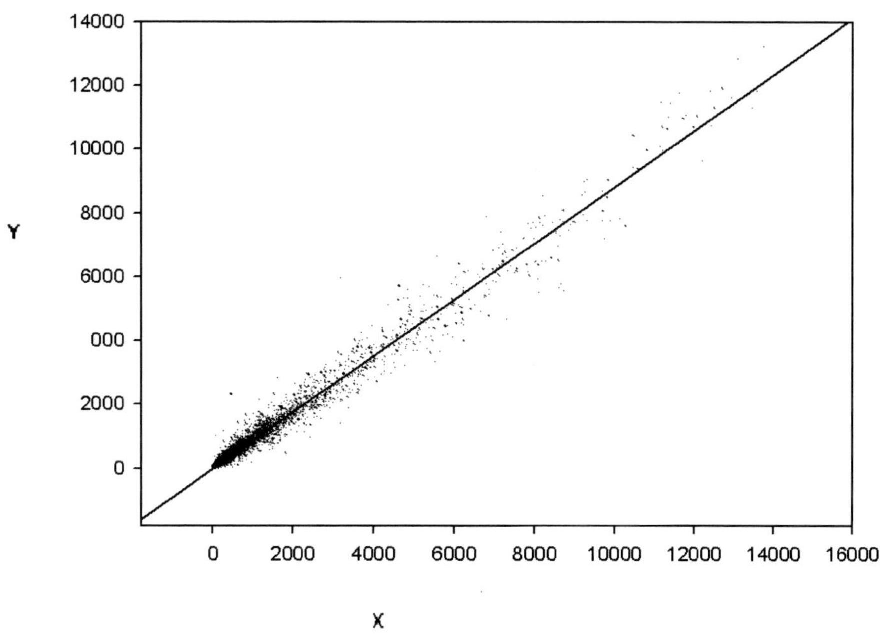

Figure 6.

The subsequent graph (Figure 6) gives a pictorial representation of this “common portion of information” for two measures (the X and the Y of the figure): the shared information is the line crossing the X,Y plane, while the deviations of the experimental points from this line correspond to the peculiarities of X and Y measures

The extraction of the analogue of the “commonality line” sketched above for a 13 dimensional plane is safely executed by reliable computer software. What is important to know is the intuitive meaning of this “shared information” which is depicted by the two dimensional picture.

This shared information has an explicit formulation in terms of a linear function of the original measures, i.e., the first principal component, the cumulative index of complexity collecting the information shared by the 13 theoretical indexes has the form:

$$PC1 = a*T1 + b*T2 + c*T3 + ...m*T13$$

with a,b, .., m being scalar numbers and T1 to T13 the 13 complexity scores.

The PC1 score can be immediately computed for all the studied series. This score gave the expected ordering to the standard series as depicted in Figure 7, where the concordance of PC1 (cumulative complexity score) with the sources of order (logarithm of the length of ordered patches for ran (i) series, signal to noise ratio for snr (i) series) is striking, reaching a concordance near to maximum in terms of Pearson’s r correlation coefficient (the maximal value attainable by this coefficient measuring the concordance of two measures is r =1).

Thus we have obtained a synthetic measure of complexity, perfectly in line with our needs (ranking of series from the most predictable to the most unpredictable) as we can judge from both the Figure 7 graphs relative to standards but even from the analysis of PC1 scores for very unpredictable (radioactive decay) and predictable (sine waves) series.

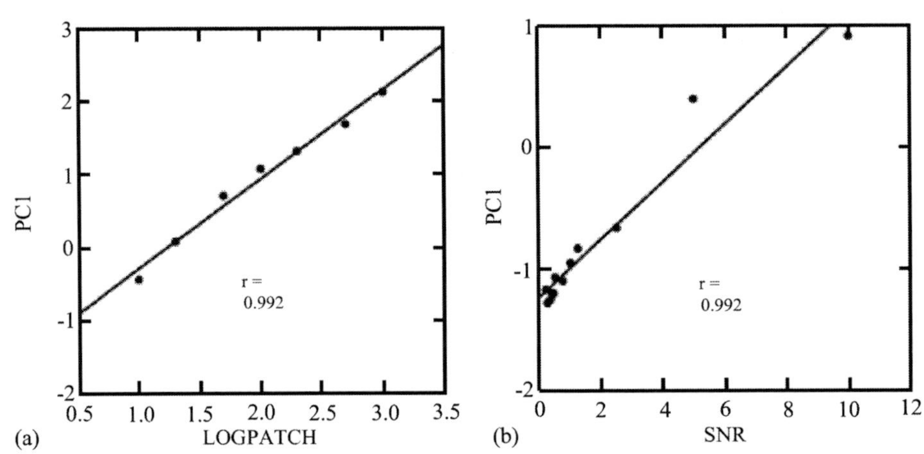

Figure 7.

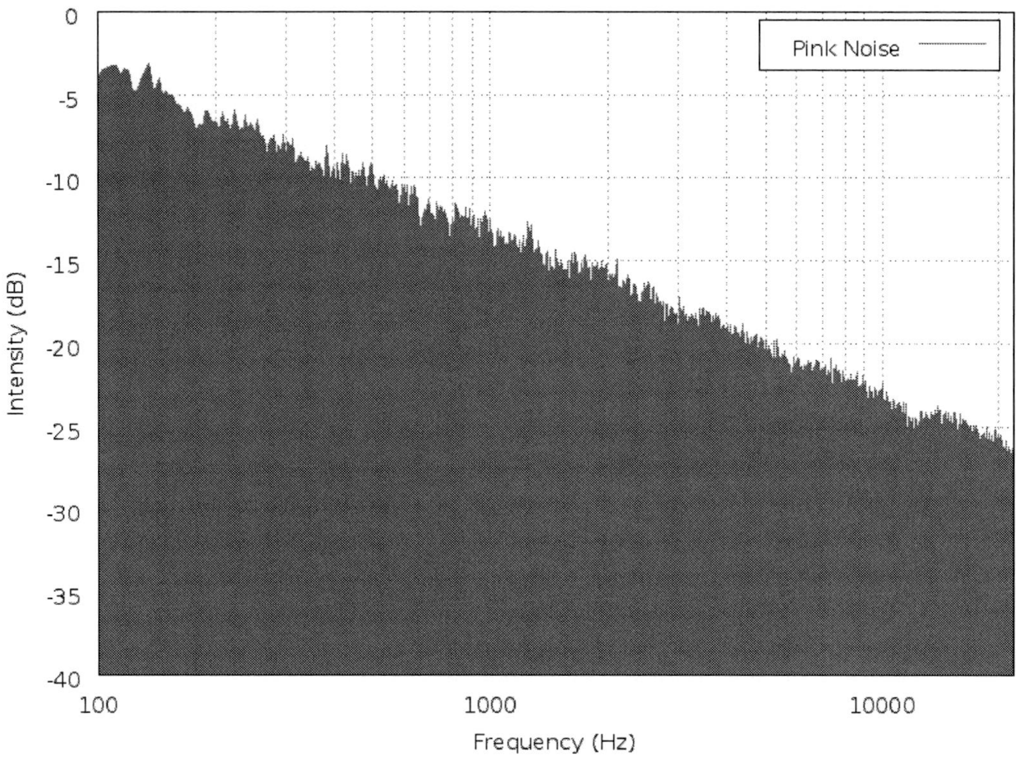

Figure 8. A Form of Noise (Pink).

Complexity passes from a vague concept to a completely operational one without the use of any *a priori* assumption. Instead, it is operationalized by the simple use of a style of reasoning typical of empirical sciences that nevertheless is useful for defining a concept of hard sciences.

To some degree this approach is a bit distressing for the hard sciences. Physics, for example, likes to reason with specific theoretical examples amenable to mathematical description. Thus one can talk about a "mass less" object, etc. And even if the experimental side must deal with some deviations, the deviations are very small. The situation is quite different for biological and especially the behavioural sciences: by their nature, biological and behavioural "observations" are more "noisy." It may be argued that these sciences lack the "precision" of the physical sciences, but this is not necessarily the case. As has been suggested, the physical sciences approach phenomena from an almost idealistic perspective, and the phenomena are tightly "controlled" in terms of their discussion and description. Biological and behavioural sciences do not have this luxury. Their "phenomena" are composed of aggregates (whose individual components may be described in physical terms at a certain level). Thus the phenomena of biological data become an ensemble "statistical physics." Behavioral sciences are further constrained by "self-awareness," in which case agent and group "freedom" and action must be understood. In sum, as one moves from the physical to the social, the nature of the observations become more "statistical." The so-called "imperfections" of observations reflect varying degrees of "randomness," for lack of a better term. Consequently, the universe may be viewed as varying degrees of determinism mixed with randomness (Figure 8). This randomness, moreover, can have different forms depending upon whether one looks at it from a static or a dynamic perspective. Where there is no "ordering" (static) the randomness is simply part of the "mix." Dynamically the randomness may be added to the process, or it may reflect the inherent variability of multi-choice behaviour.

The situation is graphically demonstrated by histograms of observations from the different scientific traditions (Figure 9). In the physical sciences, although there may be some experimental variation, the standard deviation is rather small. As one moves to biological and behavioural sciences, the standard deviations become greater. Indeed, comparing all three, the physical science distribution is almost a delta function.

Thus one can see that statistical "reasoning" is governed by the matter observed. And when statistics are used, pen and paper calculations can become daunting. Mathematics guides the basic theory, but the actual computation must be carried out by the computer.

In some sense one could make the argument that mathematics evolved because there was no computer. Math was a language for computers before they were discovered. Math is the commands—it is not a reality but rather a relational language. Moreover, even if we bring in physics with "force, mass, energy, etc." these are still abstractions—they are not realities. They describe relations which we still do not "understand" but it "works." And in some sense, this "language" may be a result of the construction of our brains. David Ruelle has made this point. The basic idea is that although the human brain can conceive of an idealist logic, it itself is not. The brain indeed works on abstractions; i.e., it cannot be burdened by details—only the broad outlines. The example is often given that we remember the general configurations of faces—but not the individual details: an artist needs a real model or photograph to recreate a "realistic" description of a scene or face. Thus again, reason is relational—among idealistic concepts. Thus, so-called "imperfections" are details—perhaps best dealt with by the machine which can "remember" them the best—the computer.

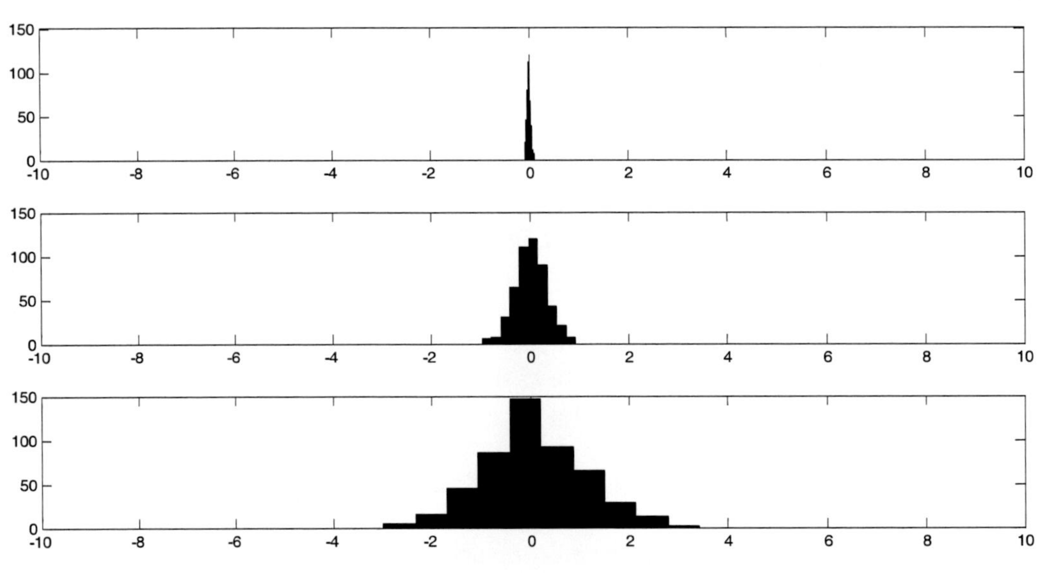

Figure 9. Histograms for physical sciences (top), biological (middle), and behavioral sciences (bottom).

REFERENCES

[1] For an extended discussion see Zbilut JP, Giuliani A (2004). Algorithmic complexity. *Encyclopedia of Nonlinear Science.* Fitzroy Dearborn.

[2] We thank Bob Eisenberg for this example. See Keller EF (2007), A clash of two cultures. *Nature* 445: 603; and Enquist BJ, Stark SC (2007), Follow Thompson's map to turn biology from a science into a Science, *Nature* 446:611.

[3] Giuliani A, Colafranceschi M, Webber Jr CL, Zbilut JP (2001). A complexity score derived from principal components analysis of nonlinear order measures. *Physica A* 301:567-588. This is the original work whose main results are reported in this chapter.

Chapter 4

THE SOCIAL TENDENTIOUSNESS OF SCIENCE

No science is immune to the infection of politics and the corruption of power.

Jacob Bronowski

One of the most disturbing statements about science is Francis Bacon's "*Scientia est Potentia*" [science is power, or knowledge is power] (Figure 1). This phrase is interesting not only for what it says, but also for the ways the Latin has come down to us. Originally, Bacon wrote in his *Sacred Meditations*, "*Nam et ipsa scienta potestas est*," [for knowledge itself is power]. The curious thing is that *potestas* appears to have been retranslated into Latin as *potentia*. The word *potestas*, is a legal term suggestive of right and authority; whereas *potentia* was often used in medieval philosophy as a potentiality (of a person or thing). Thus it suggests a kind of schism of thought developing already in the 16th century: on the one hand it is evocative of alchemy, and on the other, the political, which presages its connection with technology and war. (Interestingly, although alchemy was originally considered a neutral science in medieval society, it acquired a nefarious reputation, perhaps due to excesses and a connection with magic among some practitioners. The artistic branch was connected more to the observational, and thus less threatening.) And indeed, this interconnection perhaps could not have been avoided. Personal power is quickly subsumed by political power, and observation can quickly lead to manipulation—not always for beneficial reasons.

Although modern perception of science suggests a disinterested, objective discipline, it should be recalled that its modern origins developed at a time when religious thought still prescribed the proper view of reality. Indeed, Newton probably spent more time studying the Bible as opposed to the more "traditional" subjects like physics. Certainly some scholars feel he strongly influenced the perception of God as a disinterested creator of a rational mechanistic world. Although here too, it is not clear exactly how Newton felt about much of what he studied. He described the phenomenon of gravity, but when one examines the concept carefully, "gravity" could also be considered a magic, alchemic idea. Indeed there was nothing "physical" about gravity as he described it; rather, what Newton did was to describe it as a property mathematically existing between bodies. Certainly, an alchemist of the time would have no problem accepting the idea. To this day there are disputations regarding this concept, and even occupied Einstein's thoughts relative to "action at a distance." Indeed, contemporaneously, some related ideas or findings are spoken about as "spooky" since no easy way of explaining the phenomena are easily obtainable. Thus, one can see why science, religion and the occult still engender so many disputations. It may be argued that many political initiatives based on "rational" science (the question, of course, being the source of the rationality) have engendered ideas such as eugenics, which so horribly affected societies in the 20th century. The latent memory of such ideas may in fact influence contemporary debates about the ethics of science. Historically, concepts such as social Darwinism similarly may be seen as a natural, rational justification for national primacies and the attendant bellicose attitudes. (Is it science or political justification?) (Figure 2).

This link between knowledge and warfare has probably been present since the beginning of human culture but had a sharp transition with the development of fire arms. For the first time in history a very weak, malevolent person could prevail over a very strong, righteous one. Modern times start with this transition—after this achievement a time honored system of values abruptly changed. Many philosophers and artists of the 15th and 16th centuries (the age in which fire arms came into widespread use) clearly understood the relevance of this transition. In the ninth canto of his masterpiece, "*Orlando Furioso*," the great Renaissance Italian poet Ludovico Ariosto describes a rifle as "terrestrial lightening": [1]

Che si puo' dir che tuona e che balena
Ne' men che soglia il fulmine ove passa
Cio' che tocca arde, abbatte, apre e fracassa. (IX, 29, 6-8)

[It can be said that thunder and lights
Not differently as the lightening uses to do where it goes
What it touches fires, defeats, opens and destroys.]

Figure 1. Francis Bacon.

Figure 2. Charles Darwin.

He equates firearms to the work of the devil simply because a weak, dishonest man can win over a pure-at-heart paladin, thus placing upside down the sacred order of things. Japan's ruling class of those times had so clear a perception of this, that they purposefully banned the use of firearms on national territory so that war could remain an affair of a restricted oligarchy of brave warriors (Samurai) (Figure 4). In order to maintain this situation they completely closed their relations with Western countries. This state of affairs came to an end in the second half of 19th century, when the Meiji emperor, afraid of Western colonialism and of the consequences for national independence, decided to abruptly introduce Japan to the modern technology of war. Wars, after the introduction of firearms, stopped being a restricted affair of a social elite and became a much more bloody affair with all citizens involved.

Figure 3. Ludovico Ariosto.

Embedded in the use of firearms there is the notion of reductionism: you no longer have to care about "boundary conditions" like physical conditioning, the art of dueling, your psychological mood and motivation—the only important thing is that you center the target, that's all. A situation with many facets collapses to its center, all the accessory (but nevertheless crucial) things simply fade away and become "vanishingly small" and thus irrelevant. But this is exactly the alchemic dream of "essence!" And it is analogous to modern science's desire for "basic principles" or "laws of nature" from which all other (accessory) things can be derived by pure rational means (Figure 5). Basically, it is only our ignorance which does not allow us to do so but, at least in principle, a theory of everything that can be drawn and put into action. The world can be reduced to an equation and then human man-made science is no longer an important pursuit: all can be put in the "hands" of a computer, as has been depicted in some of Isaac Asimov's science fiction novels.

Figure 4. Samurai.

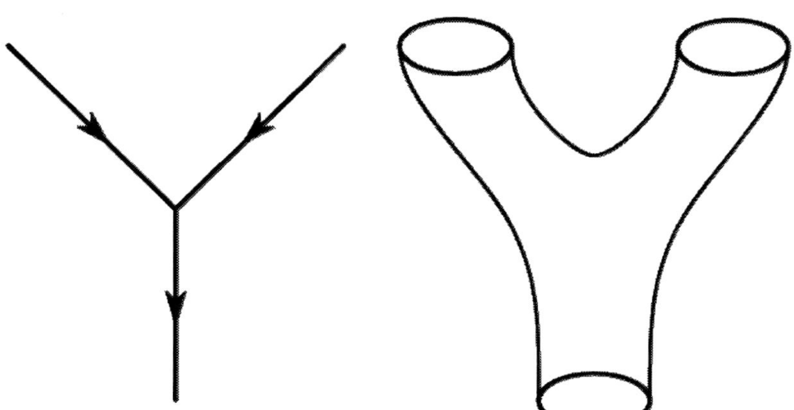

Figure 5. String Theory—A Theory of Everything.

Figure 6. Blaise Pascal.

Obviously, many thinkers of all the ages [from Pascal (Figure 6) to De Finetti] recognized the intrinsic fallacies of this way of thinking, and a significant paper by Laughlin and Pines, [R. B. Laughlin and David Pines (2000). The Theory of Everything, *PNAS* 97: 28-31] clearly and definitively (at least we hope) destroys this naïve paradigm, but this paradigm has so intrinsically embedded itself into modern science that scientific language (and consequently scientific concepts) is filled with "war-like" metaphors. Thus we can easily identify the "target" causing an illness and hitting it, like an enemy. We talk about drugs like "magic bullets," and we want to know the "exact trajectory" of a disease, like for a bomb. This kind of "warfare mentality" has had some evident successes: for example this was excellent for developing drugs against acute illnesses, and for individuating some very major etiological causes of disease such as bacteria and viruses. However, it has been less successful in the face of chronic illnesses where there is no single enemy to find and defeat, but instead many small events and co-factors—each one absolutely irrelevant by itself, but all together causing the problem.

The same is true for ecological modeling, neurological problems and practically all the real problems we must face nowadays. The persistence of bellicose paradigms for scientific discovery need to be rethought in order to regain an effective ability to operate in the real world. When describing multi-factorial "problems" one must talk of probabilities. This is the reverse of the history of "physics." In this history, we started out with major laws which are really very constrained probability distributions; whereas the probabilities on the atomic level are much more "diffuse." Thus (the De Finetti works of the thirties in this respect are enlightening [2]) we must reverse the relative role of probabilities and laws: laws are only special cases of probabilities (very constrained distributions). The root of scientific thought, from which it derives its unity, is the concept of probability which can no longer be considered as a surrogate of the "real laws to come." But if this is the case, the boundary conditions (as in face to face duels with swords) become the most important thing of all! Science comes back to its artistic branch: probability depends strictly on boundary conditions and the subjective judgement of the scientist who critically evaluates (in these two words are hidden the ethical bases of science which can no longer be considered as a necessary consequence from some premises) the evidence the scientist has in his hands is used to go forward with an experiment and judge the meaning of the results. This activity is performed by a peculiar, personal (and thus artistic, not repeatable, necessarily human and individual) use of a set of both material and conceptual tools like statistical methodology, plan of experimentation, properly assessed measures in a way not basically different from a carpenter. In the natural world, the primacy of boundary conditions with respect to basic principles can be easily recognized at the same level at which so called "complexity" starts.

Consider the contemporary problem of protein folding (Figure 7). Although genes have captured the popular mind, in fact, one thing that is ignored is that genes have a purpose; namely, to help create and direct proteins which do the major work in biological processes. One of the interesting aspects of their creation is their folding: proteins are made of unique amino acid strings (derived from genes). Because these strings are created in an aqueous environment, the strings tend to fold up directed to some degree by their propensity for exposure to water; i.e., they either like or avoid water. In principle we know the main physical forces governing this process, but in fact, this does not allow us to specifically predict the final "folding." Indeed, it is increasingly becoming acknowledged that the process can be better described as a stochastic process, i.e., a probabilistic one. The statistics of the processes involved suggest that at several important points, folding becomes a multi-valued choice. Depending upon the "boundary conditions," a protein may fold in one of several different ways. Furthermore, the processes are time-dependent. Depending upon what point in time you look at it, you may see a different kind of folding. Moreover, the "form" of the final folding brings to it varying attributes in terms of whether it is "functional" or not.

This points to the fact that the processes are "complex," but the complexity comes not from a deductive deterministic scenario, but from a combinatorial one. At each of the important junctures of the folding, the probabilities possible (and the actual "route" chosen) govern the final form. Thus one can quickly see that "complexity" in this context is relatively simple: choices are often easily identified, but their concatenation quickly becomes a "complex" problem difficult for the human mind to appreciate without the help of computers. In other words, it would seem the complexity is a question of size—specifically a size which would appear to be manageable for a human brain to grasp. Is there a specific size? Probably not. It would appear to depend upon age, maturation and prior education/experience.

If we think about, for example, an average adult's ability to understand how electricity is directed to an home, in most cases, it would probably seem to be complex. This is why, if there is an electrical problem, the typical homeowner calls for an expert, the electrician. To the electrician, the system is relatively simple. Which points to the observation of anthropologists that as cultures enlarge with divisions of labor, they become more "complex," to the point that specialized knowledge of experts becomes an important aspect of survival. Or consider the experience of the fall of the Roman Empire: cultural artefacts as simple as sewer systems, were lost to Europe for many hundreds of years (Figure 8).

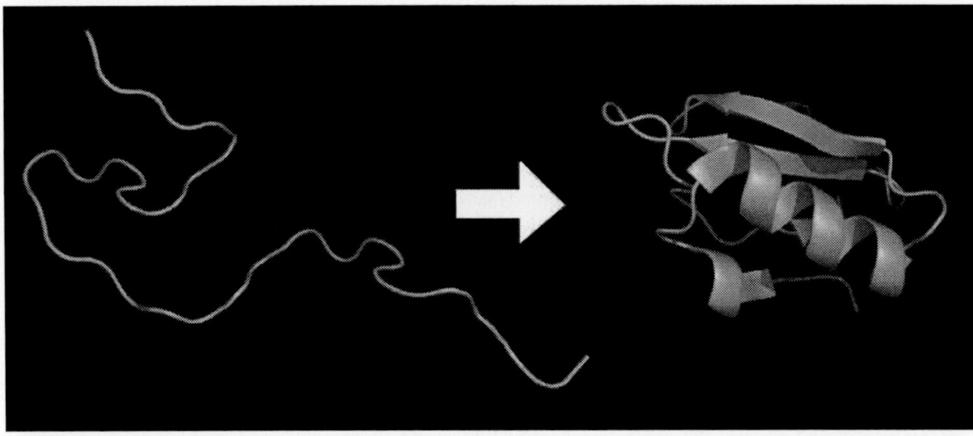

Figure 7. Protein Folding.

Figure 8. Outlet of the Cloaca (sewer) Maxima.

Even more striking is the evidence derived from the most basic of atomic processes. Consider the fact that the combination of two hydrogen atoms with one oxygen atom results in a substance (water) which has properties not resembling the original two gases. This point needs to be underscored given that a frequent remark is that complex systems are responsible for “emergent” properties. In fact, water is not a “complex system if you look at it from a constitutive view. (One may argue that in fact its dynamics is complex, but here we are speaking about the constitutive elements only.) Certainly, water is not the only example, as any basic student of chemistry knows. Or consider the periodic table of elements. Elements are distinguished by the addition of protons, yet the resulting properties cannot be *a priori* deduced from knowledge of their number of protons. These are emergent properties not obvious based on a description of their basic constitutive factors. Indeed, physicists posit these facts as a given, and have sought to find additional constitutive factors in atoms, resulting in a myriad of sub-atomic particles (and the search for a "Theory of Everything").

From what we said above it is clear that there is significant work left to methodology for deriving "holistic ways of thinking" to approach complex systems without dissecting it into pieces. We need methods that allow us to maintain a systematic, holistic view of the phenomenon even when necessarily going down to its constituent elements. (Indeed as we see with the periodic table, the constitutive elements do not in themselves provide any magical insight.) In the previous chapter we described the application of principal component analysis which is one of these tools good for this artistic approach.

Clearly PCA is not the only method to do this, and many other methodological approaches can be used as well. The point is that all these methods need "first-hand" experience of the system studied. This need to change route and cope with complexity is currently widely recognized through the flourishing of "systems biology studies," and "network paradigms." Moreover, it is a proof of this attitude. The trend seems to be the creation of "multidisciplinary teams" collectively able to look at the same problems from different perspectives. Our sense of the situation is that these teams cannot per se provide the right answer to start a new "tendentious free" science able to deal with complex problems. Multidisciplinary teams implicitly maintain the fact that any unit in the team keeps its unique disciplinary persona, and the integration is expected to come up by the simple addition of different competencies. This addition cannot spontaneously "coalesce" if the participants do not make the conscious effort to forget a great part of their special skills while retaining only those portions of operative knowledge that can be effectively shared with the other members. In our experience this can be obtained only if the team is kept together by very strong and effective friendship links.

It should be recalled that many of the great scientists of the past were polymaths—they were conversant with what we now categorize as different disciplines. Like DaVinci, many were even schooled in the arts as well as the sciences. Thus it can be said they were truly systems scientists. Of course, currently, knowledge has grown to such an extent that many scientists know only a relatively small aspect of the larger scientific endeavour. Even to this day, we marvel at a "Renaissance man," who feels comfortable in many different areas of learning. The poignancy of the current situation is exemplified in a short story by R. Macauley published in the *Kenyon Review* in 1956, entitled "The Chevigny Man." In the tale, an academic bases his whole professional career on becoming an expert about the literary output of a famous author—only to find out that his pronouncements were in effect trivial. It has become the norm that scientists must fashion for themselves a niche of exploitable expertise—at the expense of becoming truly stirred by the interests of colleagues. It is a requisite built on expediency, competition and a pragmatism for survival. Institutions of learning are now aware of the limitations of the narrow perspective and are not only creating "interdisciplinary" institutes, but also developing architectural stimulants to foster contact among diverse disciplines. But is this sufficient? Can casual contact foster a new culture? The drive for the "bottom line" can be overriding, and looking for quick results, whereas relationships need time to build trust.

The purpose is not to indict scientists. The motivations derive from institutional sponsors who increasingly demand notoriety and money. Frequently, the sponsors are academic institutions that have increasingly adopted a mercantile and perhaps political model. As history has shown us, this could be risky—not only for scientists, but also for their institutions. Science is now more often than not a costly enterprise, and institutions are always at risk of taking a big gamble by investing large sums of money in a specialty, as well as the people involved in it. The next technique, the next new discovery is always around the corner threatening to make today's cutting edge, tomorrow's obsolete equipment or someone's discredited idea.

This is not to say that the work of the single scientist is over. Too often it is assumed that collaborative science must be communal science. Although history would like to present the idea that many past scientists were solitary geniuses, in fact, they made themselves "available" to the greater scientific milieu. Even Einstein owes a debt to his wife and her knowledge of tensor mathematics. And clearly Newton could not have formulated his laws of motion without Galileo. The trick is to understand that science is at once solitary and social, and requires the skills of our species to survive in a collective way, building upon the unique perspectives of the individual.

REFERENCES

[1] Ludovico Ariosto (1516) "Orlando Furioso" is one of the three or four most famous Italian literature masterpieces, probably the only Italian piece of literature (together with Dante's "Divina Commedia") equally appreciated by the very literate and the completely illiterate thru the centuries (it is the material, among other things, of Sicilian traditional puppet theatre called "Opera dei Pupi"). A pdf version can be found at http://www.letteraturaitaliana.net/pdf/Volume_4/t325.pdf

[2] Cifarelli DM, Ragazzini E (1996) De Finetti's contribution to probability and statistics. *Stat. Sc* 11: 253-282. In this very clear work the authors summarize the main contributions of Bruno De Finetti to the edification of a probability based idea of science as opposed to deterministic paradigms.

Chapter 5

RIGOROUS SCIENCE?

> *Although this may seem a paradox, all exact science is dominated by the idea of approximation.*
>
> Bertrand Russell

A common dictum of science is that it should be rigorous. Indeed, a common critique of poorly performed science is that it "lacks rigor." Conversely, the highest accolade for scientific work is to say that it is "very rigorous."

Rigor is a Latin word with the following meanings: hardness, toughness, rigidity, freezing cold, severity, inflexibility, discipline. Thus the word evokes something (or someone) that must be kept fixed by the action of very low temperature or by the effect of a punishment, of a constraint. In the most extreme case we talk of *rigor mortis* of the rigidity acquired by corpses after a certain time after death (*mors,* gen. *mortis*). The last meaning, "discipline," highlights the need for constraint. There is the natural tendency to "rein in" erratic behavior in order to learn something (discipline derives from the Latin verb *discere* meaning "to learn"). Physics, chemistry, mathematics, biology are disciplines and they must be rigorous, thus they need to impose some hard constraints on the natural way of thinking. Why do we need to freeze something in order to know it?

At the birth of modern science an experiment was not a tool to derive some suggestions for sketching an explanation of the world—an experiment was a demonstration of the plausibility of a preconceived idea. If I believe gravitational attraction works in the same way on a heavy or a light body I must work hard to build an experimental set up that will verify this assumption. Before declaring that I am probably in error, I will adjust all the external conditions in many different ways so as to impose a "discipline" to the environmental conditions that they fulfil my desires.

This kind of vision is evident in the works of Galileo, and in his separation of "quantitative" from "qualitative" features. The former being the "ordered and ideal ones" and the second annoying accidents that mask the real underlying order of Nature. The good scientist is the one that is able to distinguish the underlying simple, and "mathematically written" ideas at the basis of Nature from the wild erratic behavior masking them. More brutally, Francis Bacon admitted Nature must be tortured to make it confess to us its truth.

Although the scientific revolution admitted a new way of approaching the understanding of reality as opposed to medieval philosophy, it yet retained the traditional dichotomy between a thing's "essence" and its "accidents," i.e., its attributes. The implication always was that reality in some way obscures what a thing really is (that is a quite funny paradox). Clearly, the idea can be traced all the way back to Plato. What is interesting is that in a very indirect way, this idea presages what modern neurocognitive studies have suggested: that the human brain retains general concepts and perceptions—not the details. We like to see patterns that apparently can be easily categorized and stored versus massive amounts of detail.

This way of thinking is very productive in the presence of a very clear and overwhelming principle (like the force of gravity) but is incredibly ineffective when scientists get stuck in an unproductive idea. Examples are replete: the search for an "ether," a fluid filling space that allows waves to travel; "phrenology," the idea you can grasp the character and the attitudes of a person by measuring some morphological features of his head and many others.

The problem is that the "real thing" was considered the scientific model, expressed in mathematical terms, and not what our senses were suggesting to us. And why not? One of the most important prehistoric, "scientific" concerns was that of what we call astronomy. Obviously, many cultures, from those of Stonehenge to the Americas demonstrated a strong interest in the movement of the sun, moon and stars devoting significant resources for the development of "observatories." The necessity of being able to predict the seasons demanded such inquiry. The rise of modern astronomy (and the mathematics it engendered) allowed for a precision not previously possible. Indeed Pope Gregory XIII (Figure 1) in 1582 established the idea of an observatory in association with the reform of the calendar. Thus there was a gradual appreciation for the fallibility of "common" sense vs. the precision of "rigorous" observation and the related mathematics.

Figure1. Pope Gregory XIII.

One cannot ignore the effect of the gradual change in the meaning of truth and reality during the medieval Scholastic period. Although Thomas Aquinas (Figure 2) is popularly considered the prime exemplar of this period, in terms of practical effect, Franciscans such as Roger Bacon (Figure 3) and especially William of Occam did more to break down the hold of theology on philosophy (and especially on "natural" philosophy, as science was called.). Perhaps most telling is the fact that the Franciscan, Duns Scotus (Figure 4) was dubbed the "subtle doctor." Effectively these Franciscans provided the breakdown of the idealist trend in the perception of reality. Their arguments, however, depended on a very subtle departure of understanding of the relationship between concepts and real perceptions (which requires its own treatise). Once the Reformation developed, it was convenient for people like Luther (Figure 5) to cite these people. Yet we argue that the two schools of thought were amalgamated into the new science, since, especially, the notion of a metaphysics became seriously disputed (or ignored). Where previously semantic precision was paramount (Aquinas was originally trained as a lawyer.), methodology and observation took over. Somewhere in the shuffle, words and ideas were appropriated in pragmatic fashion. Thus we have the rigor of scientific work: whereas "*sententiae*" were rigorously debated, now observations were—somehow still harkening back to the notion that they were masking a universal truth. Given the disputations regarding intricate and obscure distinctions about truth, one can easily imagine that they were quickly and perhaps pragmatically ignored favoring the emphasis upon observation and method. The primacy given to theology and philosophy and its knowledge ("*scientia*"), became science as we know it.

Figure 2. Thomas Aquinas.

Figure 3. Roger Bacon.

Figure 4. Duns Scotus.

Figure 5. Martin Luther.

In any case, if something can be expressed in "rigorous" terms, i.e., by keeping aside all the erratic and unexpected behaviours the actual world is made of, we are more prone to trust it—"seeing is believing."

But, for the same reason for which we admire elegant fashion models, yet we fall in love with real women, if we want to build a real airplane we look at the trajectories of toy models in wind tunnels and use the beautiful Navier-Stokes equations of turbulence only to add charm. In fact if those equations are effective why we should need wind galleries?

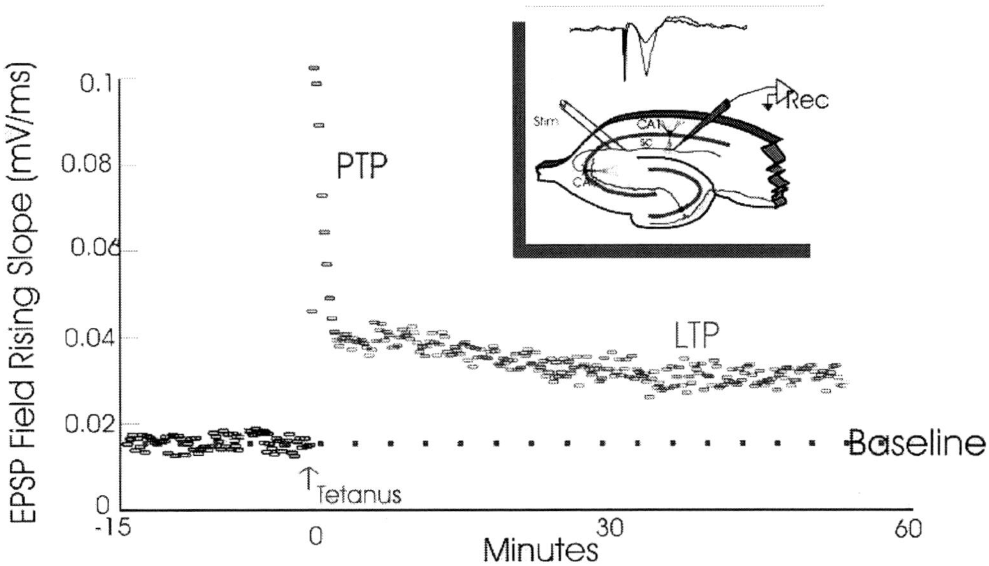

Figure 6. LTP of Rat Brain.

What is real? What is apparent? And why has reality struggled with senses? There is a joke regarding the physicist that solves the problem of predicting which horse will win a race "given the case that the horses are spherical in shape" is a clear example how ridiculous rigorous thinking can bring us. In order to be "rigorous" I must "freeze" the wild variation affecting horses down to the spherical case [1]. (Clearly, everyone knows that horses are not spherical and this is the joke's puhchline.) Navier-Stokes equations are a respected portion of science even if no airplane or sailboat obeys them. How many spherical horses are produced by the quest for rigor? How far can we go in torturing Nature?

These are very important questions, especially in biomedical sciences which are much more resistant to our imposed discipline than physics. In biomedical sciences there are many examples of pre-conceived ideas, arising by apparently plausible explanations of observed facts and for which a "crucial experiment" was actively sought for by legions of scientists, (and in many cases it seemed to be found) provoking the birth of new fields of science, funded projects, philosophical movements, brilliant careers, just to arrive at a "silent death."

A paradigmatic case is the story of the so-called "Long Term Potentiation" (LTP) (Figure 6) that represented one of the most "hot" themes of neuroscience at the end of the last century. Basically, scientists discovered that the response of isolated neurons displayed some typical features of the so-called "Hebbian" learning: they changed their response after habituation with a potentiation of the response strength when the stimulus was presented more and more in time. In other words the neurons could "learn" and this learning process can be directly appreciated at the molecular level both with the measurement of electrical transmission, and with the activation of a specific receptor [*N*-methyl-d-aspartate (NMDA) receptor complex]. This type of receptor-endowed long-term potentiation with Hebbian characteristics, allowed electrical events at the postsynaptic membrane to be transduced into chemical signals which, in turn, are thought to activate both pre- and postsynaptic mechanisms to generate a persistent increase in synaptic strength. This phenomenon was more or less immediately correlated with high-level behavioural observations like the performance of rats in a Morris water maze (a test for measuring the relative orientation learning of animals) thus fulfilling a very old dream of the scientific community: subtracting intelligence from the vagaries of "humanistic" fields and giving it a structural objective correlate. LTP soon became "the real thing." After many years of contrasting results, difficulties in reproducing the phenomenon, the increasing need for more and more abstruse reasonings to reconcile all the evidences into a common frame, LTP, sadly, abandoned the scene of the most prestigious scientific journals to hide in minor publications. Now someone still remembers it (like those old actresses that still receive letters from their fans) in commenting on some results, but the role of the star is now for other players.

Why should ideas be more reliable than evidence? This is not a crazy attitude, on the contrary, this is absolutely rational and very effective: we cannot effectively cope with the whole ensemble of information that comes from the external world—we need some generalization if we want to effectively approach the world outside us without being defeated. Thus the baseball player makes (even if unconsciously) a model of the trajectory of the ball he must catch and that model is the reality for him. If a sudden wind changes the trajectory this is considered an "error"—clearly the player maintains his model trajectory and he is totally right in doing so. What differentiates the baseball player and the scientist (and what makes the baseball player more rational than the scientist) is that the player knows that the wind exists and does not pretend to eliminate wind from the earth—he simply accepts his model is not perfect and continues to use it. Instead scientists in many cases try to discard as "apparent" (and thus not real, illusory, not existing) all the things they cannot force into their models. This attitude is obviously very dangerous for the progress of science.

A relatively recent example of such a history is recalled by Barry Marshall's winning of the Nobel prize for Medicine in 2005. In the early 1980's Marshall and Warren identified H. pylori (Figure 7) as a bug potentially responsible for most cases of stomach ulcers. The prevailing theories for the etiologies of ulcers, however, were related to stresses or other psychosocial factors. The idea that an infection was responsible went totally contrary to accepted dogma. It was in fact almost impossible for these two scientists to even present this as a suggestion at any meeting or publication. It took roughly another decade before their findings were accepted. Instead of living with a potentially chronic condition which required vigilance, patients were now given the opportunity to eradicate the infection with a short, two-week course of antibiotics. This scenario would suggest that rigor is not always objective. Marshall's proposition threatened decades of research which supported alternative etiologies; although deficiencies in the hypotheses were well known. Furthermore, significant amounts of capital were invested in pharmaceuticals which were aimed at blocking the release of hydrochloric acid in the stomach. Thus the simple finding of infection met what might be termed the "social context" of rigor. Or to paraphrase Orwell, all scientists are rigorous, but some are more rigorous than others.

At this point a suspicion is probably arising in the reader's mind; namely, the suspicion he is reading a blasphemous book trying to insinuate that rigor is not a good thing. The reader is right! Rigor is not necessarily a good thing and, in the old meaning of the supremacy of the ideal case over the observed one, it is becoming a very dangerous principle for the advancement of human knowledge when studying complex systems. Clearly, notwithstanding this, technology goes ahead: we use drugs about which we do not know how and why they work. The mechanisms of action are "just so" stories pharmacologists invent for patent needs—consider the group of drugs called "statins." They have been shown to reduce the probability of having heart attacks through cholesterol reduction; however, increasing evidence indicates they act in unforeseen ways also—what is the primary effect? We have genetically modified organisms in which a gene supposed to protect from a pest, dramatically alters nitrogen metabolism of the plant. We drive cars that are aerodynamically optimised by purely empirical observations and so forth. Probably the only technological achievement for which theoretical knowledge preceded practical application was the atomic bomb, and it is not exactly something to be proud of.

The problem is that we are approaching a dead end: thus we produce fewer and fewer really innovative drugs, genetically modified soy that dramatically fails in tropical environments, and cars, even if very aerodynamic and consuming much less oil than some years ago, still need non renewable and polluting fuel sources. We need a drastically new start, and this new start, in our opinion, passes through a complete revolution in the way we look at the world, and this encompasses the choice of less "rigid" and more "warm" principles for the definition of "good work rules" as opposed to the old rigor.

Clearly we cannot pretend to have an infinitely wide eye open to the world view, without the need of any simplifying assumption or any "discipline." Neither can we accept a nihilistic "anything goes" approach putting on the same plane any kind of information source. This exacerbated relativism is devoid of any hope of success. Probably the key word here is "perspective."

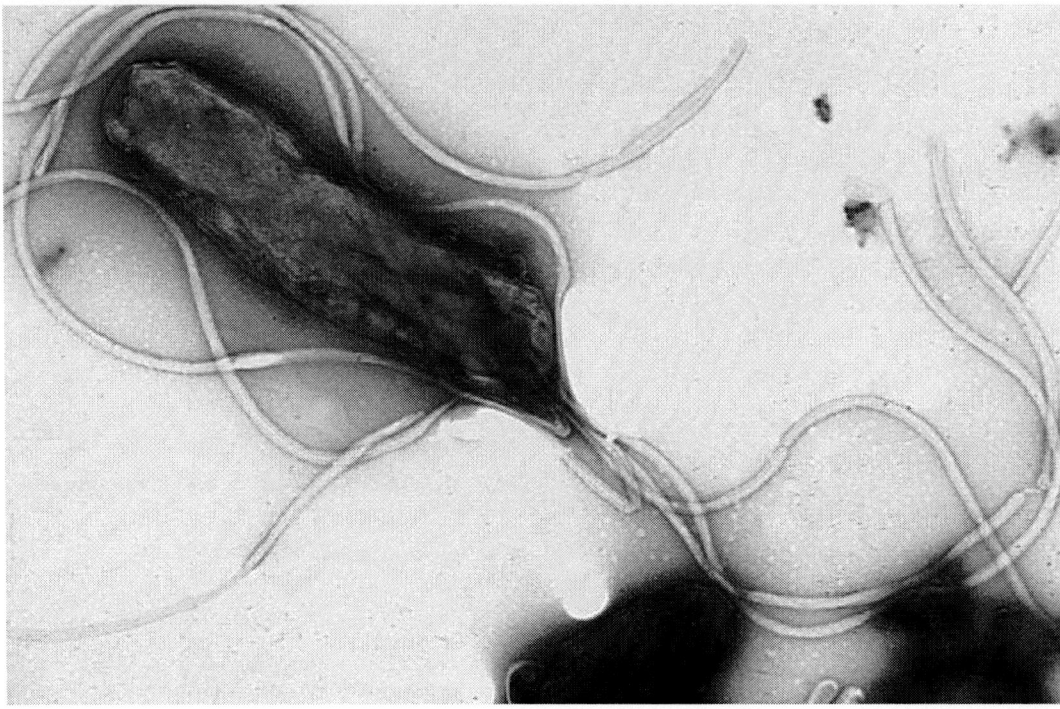

Figure 7. H. pylori.

We have a word with a Latin origin, which comes from the verb *perspiciere* and is a composite verb made by the prefix *per* (through) and the simple verb *-spiciere* (to look, whose verbal noun is *spectum*). It means "looking through" but retains the sense of something that is not only visible to one observer but can be displayed to all (this is the reason why words like spectacle or spectator come from this root). If one is said to be perspicacious, i.e. having the sense of perspective, he is perceptive, keen and clear sighted.

The art and practice of perspective was initiated and taken to an high level of advancement in the 15th century by masters like Brunelleschi (Figure 8), Masaccio and Piero della Francesca. The basic point was to define a privileged point of view of the scene represented in the painting and to accommodate the relative positions and dimensions of the represented objects so as to give an illusion of tridimensionality. The observer looks at the scene as he is located in a (predominantly but not necessarily) central position with respect to the picture. From this vantage point the laws of perspective give him the illusion of looking at a real 3D image.

The rules for the transposition of a 3D image into a bidimensional picture come from the application of Euclidean geometry. Figure 9) reports the projection of the Baptistery of Florence made by Filippo Brunelleschi. This kind of theoretical representation had an experimental counterpart: the painter "looks through" (*perspicit*) an hole made on the surface of a box with a mirror inside in which the exact bidimensional perspective of the 3D object is reported and copies the image on the picture.

Figure 8. Brunelleschi.

This corresponds to the choice of a particular measure (the hole in the box, the particular point of view) by means of which the scientist looks at the world. Each point of view is DELIBERATELY limited and local (like the hole). As a matter of fact each measure corresponds to a many-to-one mapping, of different states of the world constrained by the simple act of measurement, collapsing into the same value (thus different people have the same height, different proteins have the same number of amino acid residues, different molecules have the same molecular weight, etc.). Adopting a correct perspective corresponds to the choice of the point of view that allows for the most meaningful representation of the system at hand. The art of the scientist both clearly defines his point of view and develops its explanation of what he observed—deriving them from the features of his privileged vantage point. Thus if I am able to predict the melting point of a molecule by means of a topological index based on the structural formula, this means that the topological index (the hole through which I look at the molecule) allows for a good perspective of the melting process. By no means does this imply that the topological index is the cause of the melting, nor that we can forget any aspect different from the structural formula to explain the melting process.

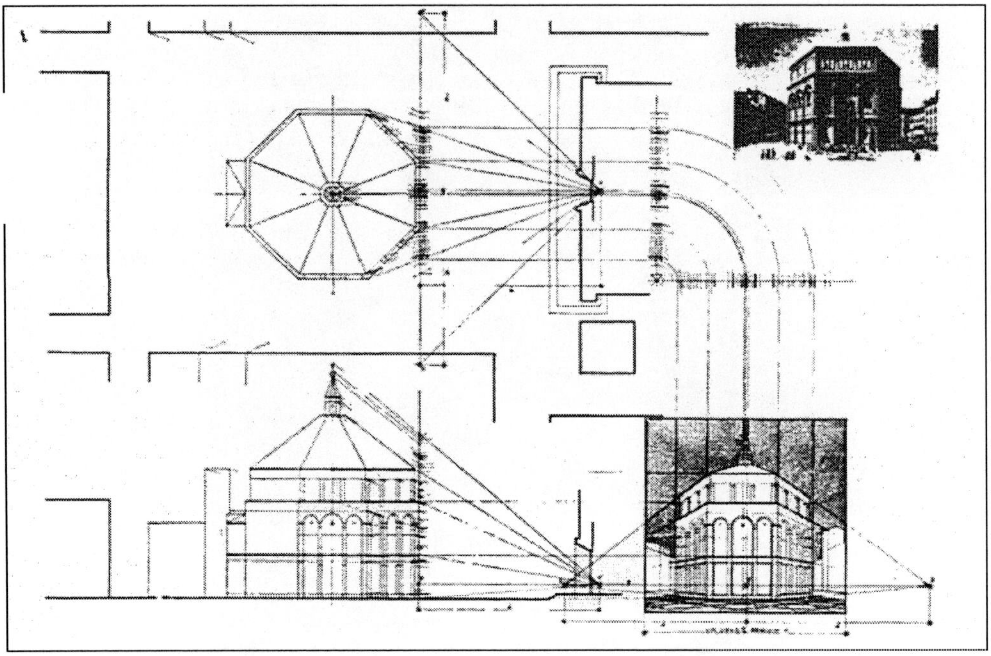

Figure 9. Perspective of Baptistery in Florence.

The generation of a good perspective can arise from the combination of a multitude of different measures (like in our example of the generation of a consensus complexity measure) but again the metaphor is still alive given the point of view is unique and clearly explained. "…the complexity measure is the consensus score of all the different applied measures." This allows for a good representation of the system because it puts in "correct" order the studied series. Shifting from rigor to perspective we exchange theory for methodology and reality for representation—is this good?

To ask this question is to return to the philosophical disputations of the medieval ages: this assumes that there is a "really real" as opposed to an apparent real. Instead it might be useful to recall one of the most useful findings of neuroscientists of the past five decades: that mental representation of the world is (in the traditional medieval sense) imperfect. The brain appears to grasp the world as outlines. The danger, of course, is that either the patterns are false (there is no real correspondence with reality), or the patterns are imperfect.

This is illustrated by the fact that in criminal trials, it has often been found that so-called "eye-witnesses" have been incorrect in the identifications of perpetrators. Nowhere more dramatically has this been shown than in the numerous cases where rape victims' positive identifications of their attackers have been found to be contradicted by DNA evidence. It was long held that visual identification of a person present at a crime was one of the strongest forms of evidence. Yet, as increasing experience based on experimental psychology has shown, this very often is not the case.

An even more striking example is the art of Julian Beever, whose 3D pavement drawings across Europe has created a bit of a cult following (Figure 10). In the graphic one can see that passers-by are avoiding the "hole" even though they clearly understand that it is not "real." The perception causes such a relatively strong emotional component that it suggests that it informs the more "rational" brain to avoid it –perhaps "just in case."

Figure 10. Drawing by Beever.

The Latin maxim states "Nemo dat quod non habit." [You can't give what you don't have.]. Similarly if our perceptual and mental machinery cannot reliably provide "the real," why assume that there is a "really real?" Reality is a constant vigilance of the mind which comes to the understanding that our minds have limitations. By constantly refining and verifying our perceptions of the world, do we understand it. It is a pragmatic venture: for a long time, mankind thought the earth was the center of the universe, and it was only with better measurements, and the resultant discrepancies did we learn this was not so. The telescope allowed us to extend our eyes to verify this. Thus the really real is not necessarily an ontological abstraction, but rather a pragmatic process which through technology extends our perspectives and allows us to see things never previously witnessed. And by seeing these things, we can learn to manipulate them.

A corollary to existentialist philosophy is that "If I do not know it exists, it does not exist for me." Although it suggests an existentialist theme, paradoxically it can also be viewed as being pragmatic. The medieval world did not know germs and viruses existed and were the cause for much disease and suffering. Practically, it meant this universe did not exist for them, and instead posited the cause for their suffering was perhaps a witch or was a punishment from God for transgressions. Unfortunately, the lack of knowledge had a practical effect.

Perhaps what science tries to do is to constantly test and explore the world to see if there are things which do not seem to "exist"—but actually do. Most interestingly is the fact that many of these things are based on scales of perception that are not our norm. Thus the macro and micro length scales often provide us with surprises. Technology has allowed us to believe in very tiny organisms, as well as very far distant galaxies.

Perhaps Thomas Aquinas should have taken his own advice as he is reputed to have said, "hominem unius libri timeo" [I fear the man of one book.]. We all know that a single perspective can give us a false impression of how things look. Looking at things from several perspectives can unveil new insights. We mean this implicitly when we are asked to look at things from several different "angles." Rigor can provide a very good perspective—but not necessarily every perspective.

REFERENCES

[1] Giuliani A, Zbilut JP (1998). The relevance of physical and mathematical modes of thought on complex systems behavior in biological systems. *Complexity* 3(5): 23-24.

Chapter 6

PERSPECTIVES

The best model of a cat is another cat or, better, the cat itself.

N. Wiener (Figure 1)

In the previous chapter we made the proposal to use the metaphor of perspective to judge science. In order to go from an obscure and, as such, practically useless metaphor to an operative principle we will describe a scientific work in which this method of doing science is put into practice.

The case refers to a paper published in 2005 [Bender A. and Glen R.C. (2005) *J Chem Inf Model*, 45: 1369-1375], the field is medicinal chemistry and the problem is to try and predict, based on chemico-physical information relative to a set of compounds, which of them will be an efficient ligand of a given biological receptor. We have two terms in the question: on one side (compounds) a set of small organic molecules about which we can derive a large amount of information by applying the vast theory of physical organic chemistry starting from the structural formula; on the other side (receptor) we have a complex biological polymer, a protein, of which we know very well the crystallographic structure and sequence. The task of trying to predict which compounds will best fit into the ligand is repeated for 11 different receptors, and for a variable number of compounds according to the following table on the next page.

Figure 1. N. Wiener.

Activity	Number of compounds in the data sets
5HT3 antagonists	752
5HT1A agonists	827
5HT reuptake inhibitors	359
D2 antagonists	395
Renin inhibitors	1130
Angiotensin II AT1 antagonists	943
Thrombin inhibitors	803
Substance P inhibitors	1246
HIV protease inhibitors	750
Cyclooxigenase inhibitors	636
Protein kinase C inhibitors	452

For each of the above activities, the scientists had a "template" compound, i.e., a drug already known to exhibit the desired activity that could be used as a model to look for similar compounds in different data sets.

The scientists approached the task with nine different methods, each one based on different assumptions about what "similarity" between two compounds is, and about what is the "relevant information" for each compound to predict its interaction with a given receptor. In other words, each method is like a competitor of a race, the method that wins is the one that predicts, with maximal efficiency, the compounds inside each data set that will actually display the corresponding activity. Figure 2 reports the predictions made by two methods (m1 and m6) together with the template molecule for one of these activities.

The nine methods went from a brute force, coarse grained one (simple atom count—the probability for a molecule to be active depends only on the similarity of the number of different kinds of atoms in the molecule with respect to the template) to a very sophisticated one, taking into consideration not only the structural formula of small organic molecules, but even the crystallographic structure of the receptor protein (DOCKSIM, a simulation of docking).

There is another very special competitor participating in the race, a "random extractor," i.e., a computer program that simply picks by chance the molecules. This is a completely dull but very important participant, because it allows us to judge the relative performance of each method by comparing it with the random extractor.

The relative performance of each method is measured by their average hit rate, i.e., the percentage of compounds predicted to be active that in fact were experimentally demonstrated to interact with the receptor, averaged over the 11 different tasks (activities). Figure 3 reports the results of the competition:

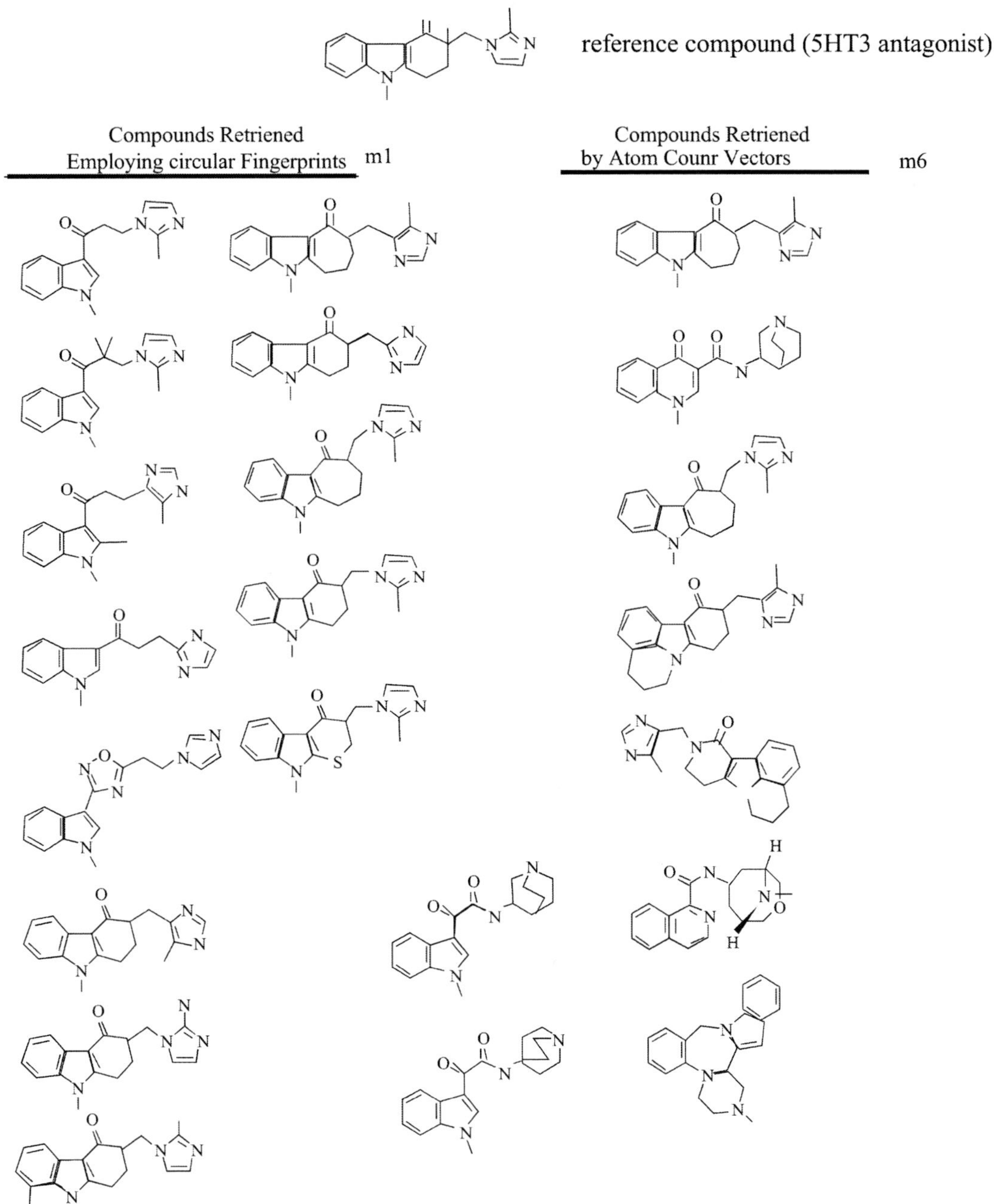

Figure 2.

The most sophisticated method (DOCKSIM) turned out to be the worst! In fact it is the most similar to random selection that, as expected, was at the bottom ranking (thus implying that all the other methods had inside some, even if small, elements of realism). The best method turned out to be MOLPRINT 2D with a very good average hit rate of 75%. The fact the "race" included thousands of compounds allows us to consider these results seriously.

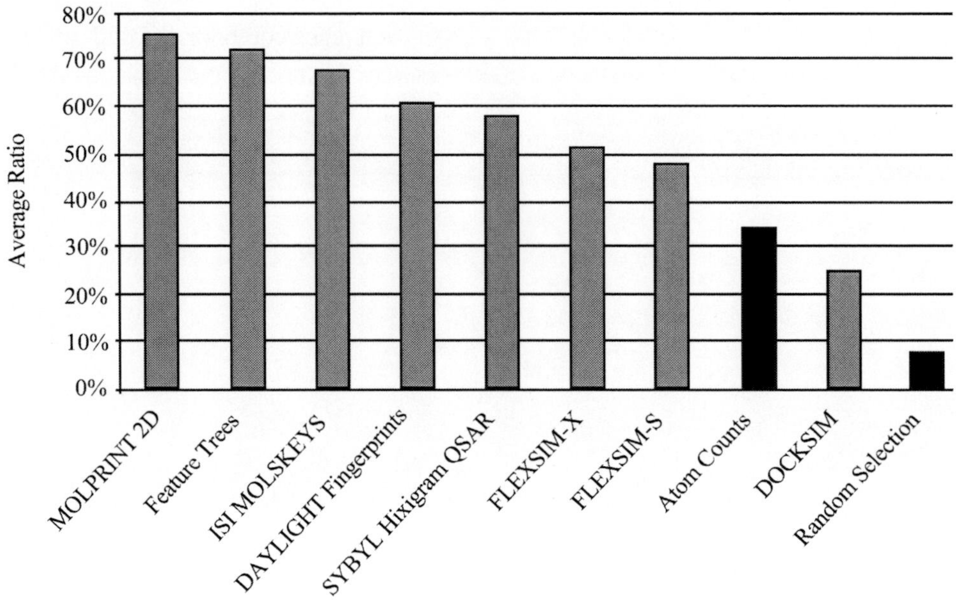

Figure 3.

Let's give a closer look at the winner: the 2D structural formula (Figure 2, the usual way in which we draw organic molecules for almost two centuries) symbolically represents certain common groups or subgroups, as well as chemical bonds. This appears at first sight to be a very poor competitor—come on! Anyone knows that the structural formula is a pure abstraction, whereas the molecules are 3D and not 2D objects, and taking into consideration (by very refined and absolutely correct scientific reasoning) this fact, can only improve performance. Yes, this is right, but DOCKSIM which not only took into consideration the real dimensionality of compounds but even the structure of the receptor (a much more sophisticated perspective) was the worst of the rational (?) methods—even worse than trivial Atom Counts.

Well, it could be that for some reason we still do not know the specific method used by MOLPRINT 2D, which was able to keep the important—stop it! All the top four competitors were based on 2D representations. Sybil (the fifth in rank) further complicated the 2D representation with chemico-physical considerations, the two FLEX# methods were both 3D (and constantly worse than 2D), DOCKSIM was by far the method most sophisticated.

We need to look carefully at these results—the case is incredibly interesting: what makes the structural formula (as expressed by MOLPRINT 2D) such a fruitful perspective? Why is it better not to be too sophisticated? What differentiates reality to allow scientists to define "realistic?"

One of the elements that make the case so instructive is the fact it is located on an interface, where "hard" science meets "soft" science. Quantum chemistry is a very developed science, making use of very sophisticated concepts and theories and allowing us to derive a world of things about a single molecule. But this is not the case for chemico-biological interactions where a small organic molecule must find its way into a groove of a huge biological macromolecule (do you remember the Darth Vader spaceship, the Death Star, of the "Star Wars" saga?). If we try and simply enlarge our perspective by means of approximations from the small to the large, we will fill our representation with more and more errors, thus disrupting the relevant information that was present in the simpler formulation. We presume to become more and more realistic by adding new elements of what we consider "real," but we do not see that we add more and more confusion to the picture. We are so superb at taking for granted that if something is scientific and more "advanced" it must work better. Instead, we simply exchange a representation (our 3D models) for the reality. It is obviously true that molecules are three-dimensional but not in the way the three dimensions are represented by our science. A bad statue is much less realistic than a good picture even if the world (and the statue) are both 3D. What makes MOLPRINT 2D such a good prediction method? A good perspective.

Let's examine a chemical formula more thoroughly, to discover that it is more than a 2D representation: first of all, it mixes symbolic (the name of atoms that implicitly go back to specific features) and geometric (the mutual position of the atoms, the chemical bonds) elements. This is a time honored and effective way of conveying information as demonstrated by many examples going from Asia to Western Medieval frescoes to modern cartoons. Considering structural formulas as simple 2D representations that must be overcome by more refined 3D representations is naïve reductionism (at best). Moreover, why insert a model of crystal structure of the receptor (like the poor DOCKSIM does) when we have already a template of a small molecule that is an effective ligand to the receptor? We could simply compare our small organic molecules to another small organic molecule (the template) that in some unknown way already has some good characteristics for doing the job. However, this seems to us too empirical, not scientific enough, and in some sense this is not totally wrong. But we must be very careful. The above medicinal chemistry example tells us how dangerous it could be to exchange our vision of the reality with reality itself, and clearly highlights the untenability of a strict linear positivistic view of advancement of science in which the "most advanced" method is necessarily the best.

A related theme is the problem of "large scale or predictive modelling" such as weather prediction. We now have perhaps three to five decent models of climate prediction which involve so much data, as well as formulas that they require supercomputers to run them. Yet how often is it that despite this, we are confronted with errors for what would appear to be a relatively simple task—the weather for the next day. Climatologists recognize this with their "error bars." Every hurricane season we get predictions of where a specific hurricane will go. With a putative "track" one gets error bounds which enlarge as the track (and time) increase.

In the past, we have argued that reality is composed of islands of determinism in a sea of randomness [1]. And it is the randomness that causes the errors. But can't we "program" randomness into such algorithms? Yes, but what is the "form" of this randomness? In many applications we use the standard normal distribution as a basis for generating the randomness, but why does Nature have to follow a standard normal distribution? (Figure 4) In fact, randomness, can mean many different things—unknown variables, mistakes in measurement, forces which are too numerous to program. And yet, we often lump them into one function. The problem, of course, is that any very small error(s) can quickly become magnified when combined in the grand ensemble, or if the process is nonlinear. Thus although one can "improve" the realism of the model by adding more information, one also increases the risk that any deficit regarding those variables can quickly overwhelm the model and provide very poor results (when compared to reduced models).

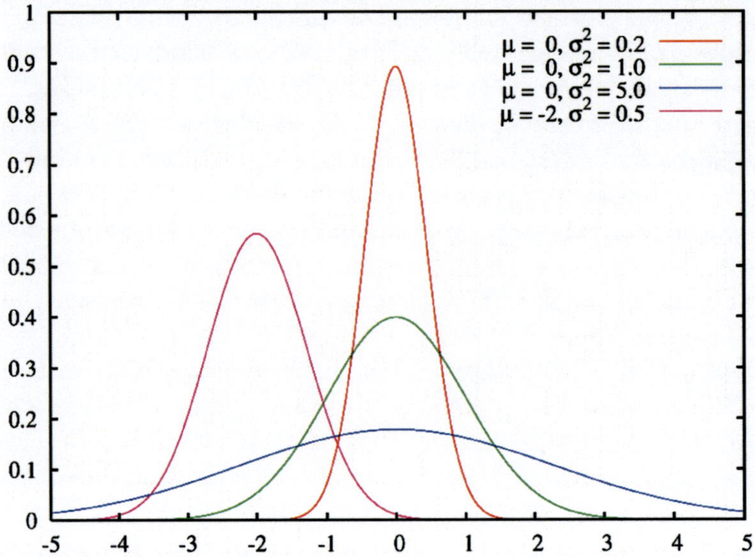

Figure 4. Several forms of the normal distribution.

The late, highly regarded scientist, Arthur Winfree, spent a good portion of his work in trying to model the electrical activity of the heart—especially in pathological circumstances. At one point, he was trying to transfer his knowledge from 2D modeling to 3D modeling. He was known to have commented that when he did this, the whole endeavour fell apart—it was as if he had to start from the beginning (personal communication).

Ultimately, the question is "How do we know when to stop?" This is the argument of a cognate pathology of "sophisticated" science, the overfitting problem which we will treat in the subsequent chapter. As for now, we have a vivid example of the "keep it simple" argument by which we started our path, and have identified good perspectives from bad ones. An important consideration is that the naïve "progressive" view stating that the best perspective is the one making use of sophisticated methods may not be the optimal choice.

Moreover, in general, we may recall the findings of Simon, that a natural human proclivity is to avoid excessive amounts of information when making a decision. Certainly, the point to be made is that when adding information, one must be sure that the information is completely correct, otherwise we increase the probability of making a mistake. One does not have to calculate a sophisticated probability to recognize that the more steps one adds to a process, the bigger the chance of failure if any one of those steps is in error.

REFERENCES

[1] Some authorities make the distinction between randomness and unpredictability, which is to say that some so-called "random" processes are really deterministic but "unpredictable." From a practical view, this may be technical semantics. For an interesting, non-specialist, unorthodox discussion, see N.N. Taleb's *Fooled by Randomness The Hidden Role of Chance in the Markets and Life, 2nd Edition.* Random House, New York, 2005.

Chapter 7

TOO PRECISE TO BE TRUE (OVERFITTING)

The most important maxim for data analysis to heed, and one which many statisticians seem to have shunned, is this: Far better an approximate answer to the right question, which is often vague, than an exact answer to the wrong question, which can always be made precise.
John Tukey

Every piece of science can be traced back to a "translation" problem: basically we observe different examples of a given phenomenon and express the different aspects (categories, intensities, etc.) of the phenomenon in terms of different values of one or more variables "Y".

If we are able to predict (with reasonable accuracy) the specific value of Y in a given instance by making use of other variables (Xs), and thus build a rule able to translate a given pattern in X into the corresponding pattern in Y, we have in our hands a piece of science. Yet, this is very formal and generic, so we cannot claim to have made a piece of science if X is simply another way of expressing Y (e.g., X corresponds to a mathematical transformation of Y, i.e., a tautology). We can say we have made an advancement if X is related to a completely different (and independent) measure we perform on the same entity of which we can know Y.

The following table reports some examples of real science:

Table 1.

X	Y	Transformation Rule
Pressure	*Volume*	*Mathematical function (hyperbole)*
Protein sequence	*Protein structure*	*Still unknown in general, some local rules for specific cases in terms of analogy*
Symptoms	*Disease*	*No general rule, case by case probabilistic reasoning.*
Number of electrons in the external orbital	*Stoichiometry of chemical reactions*	*Deterministic rules based on arithmetic*

If we look carefully at the above reported examples we immediately note that one column is lacking, namely the column reporting the specific cases in which the X/Y mappings via the transformation rules do effectively apply. We can call this column "*in the case of*" or, if we want to consider ourselves real scientists "*boundary conditions.*"

Thus the first example holds *in the case of* an ideal gas and, with an acceptable approximation (given the ideal gas simply does not exist) in a very low density gaseous environment very far from fusion temperature. The second example holds *in the case of* proteins for which one or more proteins with a sequence similarity greater than 40% whose 3D structures are already known. The third example holds *in the case of* all pathologies for which clinically accepted symptoms are described that uniquely identify a given pathological entity. The fourth example holds *in the case of* a standard reaction environment with a complete chemical characterization of the atomic species present in the reactor.

In all the reported examples, the application constraints of the rules (the science pieces) imply something different with respect to the actors playing the role of X and Y. Thus in the first case, the interplay between pressure and volume according to Boyle's law is mediated by the combined action of relative density and temperature of the gas: too many molecules break the rule because they start to interact with each other—too low a temperature makes the gas no longer a gas.

In the second example we have an "extreme" form of dependency, the ability to link the sequence of a given protein to its structure depends on other proteins! Similar reasoning can be made for the other two examples and for every scientific result. What seems to us a stringent "law of nature" has a very precarious existence living only thanks to a lot of incidental conditions that decide its applicability.

The essence of a scientist is to correctly define boundary conditions in order to establish (and clearly express to other people) the conditions of existence for discovered regularities, thus allowing other scientists to verify the findings. If a scientist is able to do this, a new theory (model in contemporary language) is born. The work of defining correct boundary conditions for an observed regularity (so to get the title of the model) currently is a major preoccupation of scientists and the argument of endless scientific literature. Let's look now at the insidious ways in which the boundary conditions issue comes into play, and to do this we'll examine an ultra-simplified model: the sketching of a linear equation able to predict the values of Y starting from another variable X. To add realism to our simulation we can think of X as a chemico-physical property of some organic molecules and Y as their relative biological activity.

In analogy with the previous chapter we can think of Y as the strength of binding to a given receptor, while X is the logarithm of the partition coefficient between octanol and water (LogP) measuring the molecule's relative hydrophobicity—a crucial property with respect to the molecule's ability to bind to a receptor.

Let's imagine we take some organic amines (basically an organic molecule having as a functional group $-NH_2$), do an experiment and obtain a result like the one depicted in Figure 1.

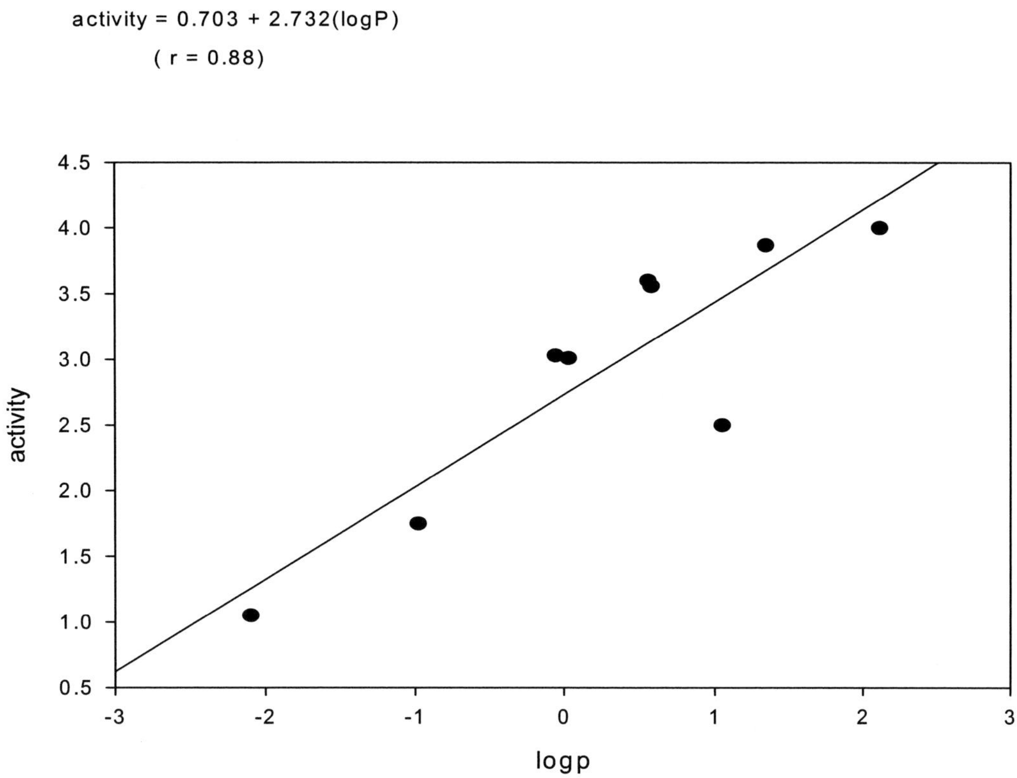

Figure 1.

The result is not bad—we obtain a significant fit between the logP value and the measured activity (Pearson r = 0.88) compatible with the presence of a normal uncertainty in the measurements. So in some sense we can say, "In the case of organic amines the biological activity positively scales with the hydrophobicity (high logP values point to hydrophobic molecules)." This is a purely descriptive statement. To test the real predictivity of this finding we must take some new molecule, compute its logP, apply the observed model [activity = 0.703 + 2.732(logP)], and check if the observed activity fits the predictions. On a pure mathematical basis the so called "domain" of the empirical function is any molecule having a logP between –2 and +2 given that this is the range of the values of X on which we perform the analysis. Clearly the model could be valid even outside this range but, rigorously (again this word) speaking, we do not know this. From a chemical point of view our domain is made of "organic amines." Obviously, the need to specify "organic amines" implies we know that biological activity CANNOT IN GENERAL BE EXPLAINED solely by hydrophobicity, but setting the boundary conditions to organic amines, i.e., to a very small subset of the entire universe of chemicals, implies that the elements of this subset can be considered more or less identical relative to other determinants of activity different from hydrophobicity so that we can sketch this simplified model.

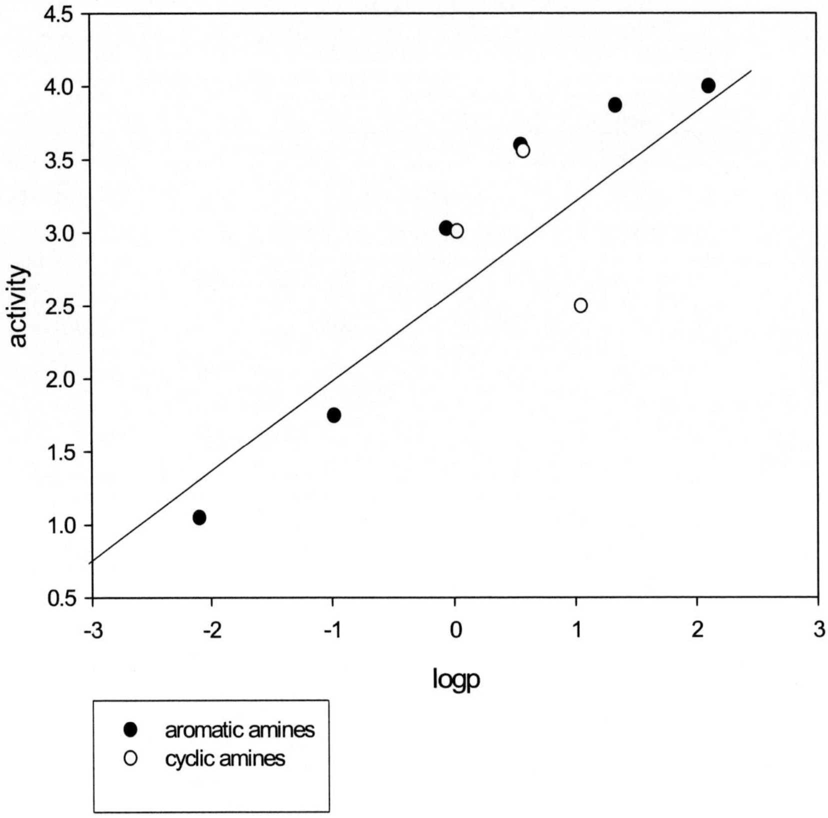

Figure 2.

This is not a minor point or a philosophical subtlety. This implies that, as in the Raymond Carver book we quoted before, with our model we are "speaking of something else," i.e., we can speak of biological activity while discussing hydrophobicity, but we know this is not the truth or better, this appears to be true simply because we are in the right perspective (the organic amines) that allow us to discard all the other (largely unknown) factors influencing activity.

Let's now imagine we test our model with some new molecules and obtain contrasting results: the activity of some molecules is very well predicted, while in some other cases the predictions are very poor. This forces us to re-think the applicability domain of our model. In this re-thinking we observe that the data set is made of organic amines, but some of them are aromatic, some others are simply cyclic with no aromatic character. The above Figure 2 is identical to the previous one but with the indication of the two sub-classes of amines.

This means that probably we have mixed two different models into only one, and this heterogeneity destroys the validity of our predictions. Consequently, we decide to perform separate models for aromatic and cyclic amines improving the statistics for cyclic amines. This gives rise to the subsequent result:

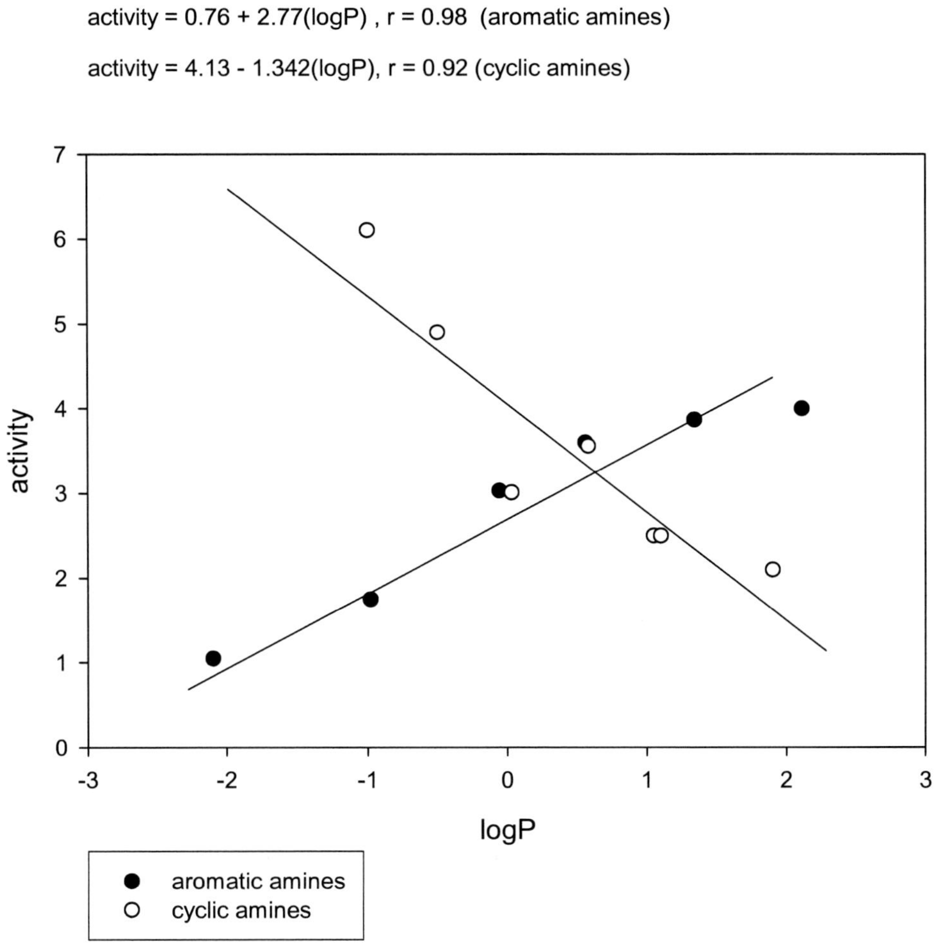

Figure 3.

As is evident in this last Figure 3 we have two different "rules" for aromatic and cyclic amines, the aromatic amines increase their biological activity at increasing logP, while the cyclic amines do exactly the contrary. This is not strange—it is simply that the two classes of molecules exert their activity with two different mechanisms. Now, though, we have the possibility to make the prediction for a given molecule by assigning to it the most suitable model of the two. It is worth noting how the fit of the two specific models (0.92 and 0.98 Pearson correlation for cyclic and aromatic respectively implies a striking precision) is better than the fit of the general one, simply because we have more homogeneous material. This is a very good result, but someone could note that, looking at the aromatic amines we could fit something more "realistic" to the data, by taking into consideration the fact that at high logP the relation between the X and the Y tend to reach a plateau and thus tend to become non-linear. This element can be taken into consideration, and we could obtain a fit like the one depicted in Figure 4.

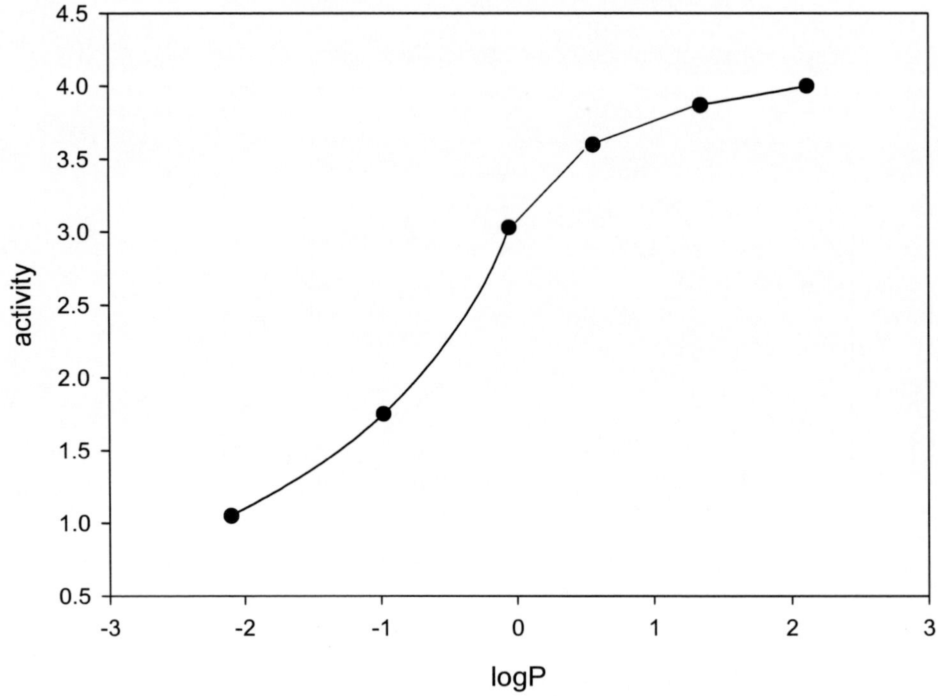

Figure 4.

Figure 5. LaPlace.

Now the fit is perfect. If in the linear case, an r=0.98 implied around 7% of unexplained variance (being the proportion of the variance explained by a model—the square of its correlation coefficient), this new fit, taking into consideration the nonlinearity of the data, reduces the uncertainty practically to 0, and allows for a perfect reconstruction of the observed activity (the points lie on the fitted curve). Nevertheless when this apparently perfect model is tested on new molecules not taken into account in the fitting procedure, it shows a worse predictivity than the simple linear model. This is a typical pathology of models—this is "overfitting."

What happened is that, with the goal of completely explaining all information present in the data set, we went into very specific peculiarities of the data that cannot be generalized to any other case, i.e., from a certain point onward we are modelling the "noise" (uncertainties of measurement) and not the empirical relation linking the studied variables. This is at the basis of the rapid decay of predictivity of models when they become too precise.

This may go contrary to the idea that the more accurate a model is, the better the predictive result. A model, by definition, implies a general outline of the most important variables necessary for the description of a phenomenon. By adding more specification (and perhaps unintentionally unmeasured noise), the implication is that all the included variables are necessary for the correct result. The point of model making is to determine what constitutes a minimal model. This implies the "demon" of the 18th century mathematician, Laplace (Figure 5), who suggested that one could precisely predict the future of the universe, if one could know all the forces of nature. A corollary to this would require a computer capable of dealing with all these variables. However, this is not the case, and a considerable debate continues on ultimate computational capability. Thus, at least for the present, we do not have this precise capability—yet models often imply we do.

If anything, a more reasonable consideration is the razor of Occam, that beings should not be multiplied without necessity. These considerations are often forgotten in the race to build a better model. There is the natural tendency to think that the more information I add to a model, the better it will be. But in point of fact, a reduced model based on given boundary conditions, is often much better than a large model whose conditions and/or measurements are poorly defined. This may especially be the case in nonlinear models, where often very small inaccuracies can propagate into the future and be significantly amplified.

In the above example we took into consideration a very simplified version of model, but think of much more sophisticated kinds of models like the ones now used for getting rid of high throughput data in molecular biology. In this case we must get rid of mutual correlations between thousands of variables (e.g. expression levels of different genes), while trying to check the consistency of what is observed with the content of huge databases reporting very intermingled connection networks among different gene activities coming from thousands and thousands of previous experiments.

We deal with extremely complex and flexible explanation models able to adapt themselves to virtually any possible experimental result. This implies that we have sufficiently rich knowledge to explain everything—or nothing. This paradoxical (and extremely dangerous for the advancement of science) situation is very well described in the paper by Rzhetsky et al. (*PNAS*, 2 103: 4940-4945, 2006) who explicitly call "microparadigms" ideas that, regardless of their verity, become more and more pervasive in the scientific community because the results they come from are no longer under scrutiny, and their consequences are simply taken as a an undoubted truth to build other models.

Figure 6. Thomas Bayes.

This is the curse of the Bayesian (Figure 6) way of reasoning when the size of the knowledge base becomes too vast and powerful.[1] This is similar to building a big tower accepting at each layer a vanishingly small amount of instability: when the layers become too many, the tower is not only globally unstable but we can no longer trace back where the error was. When simplicity and clarity (aesthetic principles, Occam's razor) are no longer guiding principles, a scientific tower of Babel may ensue.

Curiously enough modern science was born in the 17th century thanks to a "Babel-tower-like" collapse. Kepler (Figure 7) had the possibility of explaining planet orbits maintaining the old idea of circular orbits by using a combination of the so called cycles and epicycles. This kind of computation enabled astronomers to obtain perfect representations of planet trajectories.

In figure (8) the epicycles-based trajectory of the planet Mars is reported. Now we know why epicycles work well in any circumstance—they are nothing more than a different version of the Fourier spectrum of a trajectory, and there is a theorem that says that every signal, no matter how complex, can be traced back to a convolution of sine waves.

But for the same reason that this method gives in any case a solution consistent with the data, it cannot be used as an effective model—it is simply a rephrasing of what is observed using different words—it is really a tautology. Kepler made the fundamental move to come out from the sterile perfection of epicycle calculus and then, giving an important impetus to modern science, he decided to assume orbits were not circular but elliptic. This simplified the computations, and did not worsen (neither ameliorate) the quality of the predictions. It was simply an aesthetic move toward a more "beautiful" explanation that made it possible to go forward in the explanation of astronomy. Because the mathematics are simplified, it might be a case of the application of Occam's razor.

Figure 7. Kepler.

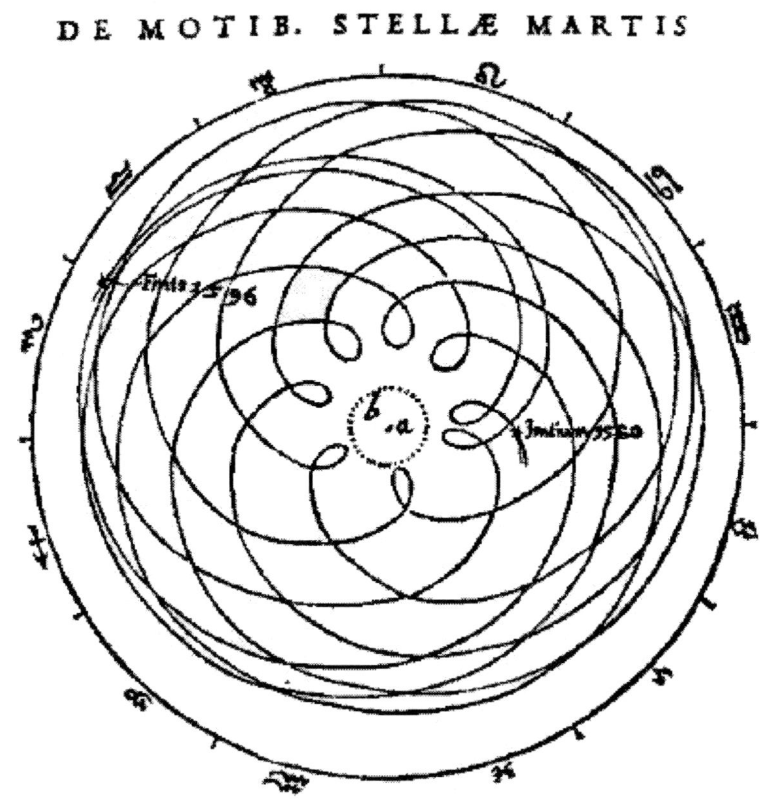

Figure 8. Geocentric Motion of Mars.

Soon or later we must dare to abandon our way of reasoning and, curiously enough, we must abandon them when they become too powerful. There is a tendency as Rzhetsky et al. have suggested, that an "information cascade" tends to inform new phenomena. When new data are explored, there is the tendency to interpret it through the perspective of previous possibly related work—irrespective of its accuracy. There is the idea that if patterns are somehow similar (but not necessarily strictly alike), the same forces are at work. Again, part of the motivation for this is the human mind's desire to see patterns—and to add more evidence confirming the similarity, rather than findings points which are not similar.

This is not a new problem. Certainly, the need to have a "control" addresses this problem, yet, in fact, an erroneous opinion can perpetuate despite evidence to the contrary, once the erroneous opinion gains a foothold. To this day, people will often suggest that a person has a "type A" personality deriving from a popular book on the subject several decades ago. The problem was that that the observations were derived from the populations cardiologists worked with—obviously predominating with patients with cardiac problems—they were highly unlikely to see patients with cardiac disease who had "type B" behaviour. Indeed, subsequent studies have strongly questioned this basic paradigm. Although there are some elements of truth to the idea, but much more explication is required to appreciate it. Yet, currently, the idea has entered into popular culture with very little discrimination. More importantly, numerous research articles accept the paradigm with very little specification. In statistical terms, the type A/B personality model may be statistically significant, but it explains very little variance. The danger with this is that efforts could be more specifically directed to treatments of subsets of populations and be more productive, as opposed to treating general personality categories with relatively small effect.

One can easily determine "type A" personality traits in almost any person—and focus in on them. For example, one may find people live will into their ninth decade, have type B features and conclude that their personalities were the result of their longevity. However, one might also find similar groups who had type A personalities. A not surprising commonality between both groups might be the fact that both groups' ancestors tended to live very long lives irrespective of their personality types; i.e., they possessed a genetic heritage which promoted their longevity. [2]

The point of the example is to underline the fact that with increasingly large amounts of data, it becomes more and more difficult to single out those distinctions which make a difference. Scientifically we speak of these distinctions as "boundary conditions," or contexts. Thus the problems associated with such researches are not so much with their "complexity," as with the size of the problem. The reason is relatively simple—if a process is composed of several independent probabilities of execution, the final probabilities are calculated as a multiplication of the probabilities. Thus the difficulty quickly grows as a combinatorial problem—both mentally and computationally.

REFERENCES

[1] This refers to the idea that "prior" probabilities need to be considered in evaluating a hypothesis. See, e.g., Berger JO (1999) *Statistical Decision Theory and Bayesian Statistics*. Second Ed, Springer Verlag, New York.

[2] The reasons for the persistent belief in the Type A/B personality are difficult to explain. One large-scale study puts strong doubts to the theory. See Pilia G, Chen WM, Scuteri A, Orru M, Albai G, Dei M, Lai S, Usala L, Lai M, Loi P, Mameli C, Vacca L, Deiana M, Masala M, Cao A, Najjar SS, Terracciano A, Nedorezov, T, Sharov A, Zonderman AB, Abecasis GR, Costa P, Lakatta E, Schlessinger D (2006) Heritability of cardiovascular and personality traits in 6,148 Sardinians. *PLoS Genetics* 2:1207-1223.

Chapter 8

Science Styles – First Elements for a Critique of Science

Science is built up of facts, as a house is built of stones; but an accumulation of facts is no more a science than a heap of stones is a house

Henri Poincaré (Figure 1)

Figure 1. Henri Poincaré.

For many human activities (even very complex and difficult to learn) some experts or "connoisseurs" exist who, while not being able to perform such an activity, are able to appreciate the different level gradations and the general value of the activity. These experts play a vital role in connecting the specific field to the rest of the society.

This is the case of art and literature experts, of movie critics, sports journalists, cuisine experts, evaluators of economic activities. The fact that these kinds of experts do not exist in science is remarkable and it is the cause of some pathologies within the scientific community, which are becoming more and more evident. The proposal of what could be the basis for the foundation of scientific criticism by laymen not directly involved in science is one of the main goals of this book and strictly relates to what we introduced as "Michael Jordan's simplicity" in the first chapter. The usefulness of such critics for science could in principle be significant: going from the most efficient allocation of resources for research to the development of truly innovative ideas for the cross fertilization of different fields.

In previous chapters we implicitly gave some elements of what we think could be the basis of such an "aesthetic" assessment of sciences by pointing attention to some critical aspects of science. Now we will try to be a bit more specific from the vantage point of our "nomadic" sojourns in science which have given us the possibility of collaborating in many diverse scientific enterprises from applied mathematics to clinical and psychological science.

Clearly, as any honest aesthetic critics, by no means do we have the intention of being general and objective. On the contrary, we are proud to state these are very personal and subjective aesthetic positions expressed with the aim (among others) to stimulate debate and, why not, to push other "experts" to disagree with our positions and to propose different lines of interpretation.

As indicated by our book title, our line of thinking can be summarized as the pursuit of "simplicity without simplification," i.e., the quest for the most clear and essential kind of reasoning (emphasizing simple models and keeping to a minimum the mathematics) while having clearly in mind that complex systems are not the best place to find definitive and all-embracing laws.

In order to sketch a modest proposal for the foundation of such a criticism, we will make use of some of the stylistic elements we introduced in the previous chapters when analysing different pieces of science, putting the accent on the procedures the authors used for their works more than on their actual content. This is the very basic principle of any tenable critique of science: it should be mainly a procedural (largely content independent) critique. This by no means implies that science is a mainly procedural affair, on the contrary its specific content is of the utmost importance, but we can largely assume that the agenda of science is dictated by interests external to science and linked to priorities set by society (including the priority of advancing basic knowledge of nature without any specific applications in sight). Thus we consider a scientist addressing a somewhat relevant question and we must simply check if his answering strategy is more or less consistent with his stated goals. This corresponds to the duties of accountants during an audit of any given firm's budgets and plans: they do not question the kind of job the firms are involved in, rather they simply check if the profile of the firm in terms of cash flow, resource allocation, financial structure, marketing strategies and so forth, are in line with the declared goal.

Figure 2. Proteus.

This kind of procedural perspective is often obscured by the classical "peer review" evaluation typical of science, where the judges are people involved in the same business as the author. Think of what could happen if movies were only watched by actors and directors or songs only listened to by musicians, and not by independent actors or musicians but actors who had wanted to be in the cast or musicians that play similar songs (in fact the reviewers that judge a specific aspect of science are strictly involved in the same kind of research). Simply, the system would become increasingly self-referential with the community becoming stuck on different topics that are considered "interesting" by the whole community (or by some influential leaders in the community), and then are exploited producing seemingly identical pieces until something else becomes "the best game in town." And the community suddenly abandons the old hit and goes for the other. This is exactly what is currently happening in science—it is a phenomenon very well known by epistemologists, called the Proteus effect, and its consequences on the efficacy of science are devastating (Figure 2). This effect is further amplified by the existence of "leading journals" dictating fashion. [1]

There is an apparent economical paradox of the business of science: when one of the authors of this book (AG) published his first paper in a peer reviewed journal he told the good news to his father who in turn asked him, "Very good! How did they pay for your article?" When it was explained that most scientific journals do not pay for articles, on the contrary, they charge a page fee for publication, the father said, "Well, this means that your articles are not so important, since anyone can have his ideas published if he pays." AG's father was not a scientist, he was a lawyer, and his way of reasoning was perfectly reasonable in the real world except in science. (By the way, the only glorious exception to this attitude we met in our experience was the prestigious *Chemical Reviews* who paid us an honorarium when we had a paper published, thus explicitly assigning a direct value to our job). Nevertheless science is part of the capitalistic world, so if scientific papers do not deserve payment but, on the contrary, journals ask the authors to pay, this means only one thing: scientific papers are advertisements (and this is what honestly some journals like the *Proceedings of National Academy of Science* state for each published paper).

Why in the world do scientific papers advertise? The answer is simple: the objects of advertisement are the authors of the paper who by publishing demonstrate to their peers they have "scientific merit," and thus they have a better chance when applying for funding. At the same time their egos are satisfied. Well, but this is not exactly the idea of science that non-scientists have. The general idea is that each piece of science is a small but different from zero step toward the advancement of knowledge and consequently, the achievement of a better life for all and so on. This mythology is so pervasive that recently a famous scientist was convicted of sexual violence in the case of a 14 year old girl, but because he was a scientist, there was a strong outcry for his pardon, suggesting he could atone for his crime by continuing his work for mankind.

Obviously, we too are convinced that scientific papers, in the long run, contribute to the advancement of knowledge but this can only be applied to a strict minority of scientific output. Moreover, the fact that knowledge contributes to the advancement of mankind does not depend on knowledge but by how it is used. Our proposal is much more modest and starts from the strict analogy between art and science we set forth in previous chapters, which is: "A relevant piece of science is a nice piece of science, independent of its future applications. If a piece of science satisfies the aesthetic attitude of the reader (listener, user) by stimulating his critical thinking and evoking new and unexpected hints, this is a good piece of science."

Clearly to be good, a piece of science must be honest, i.e., it must not give the illusion of saying more than what it can effectively say. This goal can be achieved in different fields and with different types of scientific output. Here we will enumerate three basic kinds of science with two or three tips for each type we hope can help to discriminate the masterpieces from less relevant pieces.

- *The descriptive study*. This kind of work can be equated with large frescoes decorating churches or palace ceilings and walls. Inn frescoes there are several things happening at once, and the eye is taken by a multiplicity of stimuli. Differential gene expression studies, metabolomic and proteomic studies are the classical representatives of this kind of science. Typically we are presented with thousands of different variables (genes, proteins, metabolites, interactions) redounding to a basic theme like "the discrimination of patients of a given disease from normal people" or "the characterization of a given cellular type" and so forth. It must be kept in mind that the most dangerous risk of this kind of work is overfitting: if I can use thousands of variables, I can practically discriminate anything I want by the simple action of chance. Thus for this kind of study it is of utmost importance that the authors face the specific discriminatory task only after having reduced the degrees of freedom of the system with an unsupervised strategy, i.e., without explicitly referring the statistical units to their specific class (e.g. disease/normal). The discrimination can be obtained on the basis of a very small number of degrees of freedom to keep to a minimum the chance of spurious correlations. This can be done by means of many different methods (PCA, Cluster Analysis). It is important that the solution is internally consistent without relying too heavily on external literature. It is trivial to say that gene X was already demonstrated to be important by Dr. Y. Nowadays one can find support for practically any idea that one can come up with, as in the case of Kepler epicycles. It is much better to think of splitting the statistical units into a training and a test set and then demonstrating that a parsimonious model built on the training set is able to get rid of the test set also without any further modification. As in the frescoes of the church of St. Francis in Assisi (Figure 3), all the different elements of the scene must collaborate to give the same general sensation to the observer so that the multiplicity gets organized into a single understandable message.

Figure 3. Fresco by Giotto from Church of St. Francis in Assisi.

- *The simulation study*. This is potentially the most "free and honest" kind of study of all. In principle, the scientist can prepare things in a way to be completely devoid of any inconsistency: given that the data do not come from experimentation but are numerically generated by a well explained recipe, he can eliminate *a priori* problems like overfitting, hidden variables, measurement errors. Simulation studies can be equated to abstract paintings like the works by Kandinsky (Figures 4, 5) or Pollock. The main risk here is the lack of relevance, the crucial point being the connection the scientist establishes with the reality he wants to *simulate*. This connection is based on a simplifying assumption like "The metabolism of a cell can be thought of as an electrical power grid." This is an interesting idea but we must keep in mind that metabolism is not a power grid, so we can use this simplifying assumption only up to a certain point, e.g., only when considering what happens if I eliminate an enzyme (corresponding to cutting a line in the power grid). Clearly I cannot make use of Ohm's law or establish some energy conservation law analogue to what happens in electrical circuits. These pieces of science (like abstract painting with respect to more naturalistic ones) are thus much more difficult to evaluate with respect to others because we must have a non trivial knowledge of both the "inspiring" (power grid) and "represented" (metabolism) reality to judge. For this reason we must ask the scientist to carefully explain the how and when the simulation mirrors the reality and to clearly define the limits of application of the simulation.

- *The refined measurement study*. This kind of study is the typical study that becomes fashionable when a new instrument comes into play. This was the case of functional magnetic resonance in neuroscience, of atomic force microscopy in structural biology, polymerase chain reactions in molecular biology and so forth. The problems of this kind of study are "translation" problems. These pieces of science can be equated to poetic works in a given language that can be only translated (with huge difficulties) into another language by other poets and with a huge risk of completely ruining masterpieces (this was the experience of one of the authors when reading two different Italian translations of John Donne's poems—one allowing the reader to discover an immense genius, the other frankly embarrassing). The risk in these kinds of studies is in making an unjustified parallel between the specific physical quantity measured and the conceptual entity this physical quantity is supposed to represent. Thus how could we be convinced that a spot in a brain map, linked to the amount of oxidized haemoglobin is the "image" of an on-going neuronal activity or of a "mental process?" The scales in both space and time involved in these processes are widely different and in many cases not comparable. Nevertheless we can try and find some interesting relations. The two antidotes against unjustified claims in these fields are: a) an almost maniacal attention to the statistical subtleties (are the results coming from grand averages or are they real signals? How many degrees of freedom? Which probabilistic assumptions?) and b) keeping the lowest possible profile in terms of extrapolations by considering the measures as "proxies" of the "real thing" and not the thing *per se,* at least until when a necessary link between the measure and the measured object (the chemist can think of the Lambert-Beer law for spectroscopy) is established.

Figure 4. V. Kandinsky.

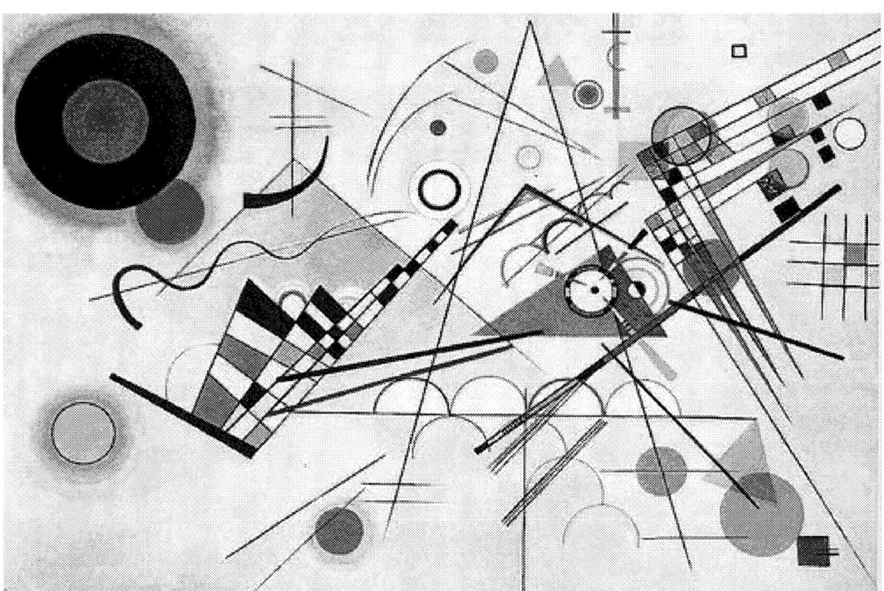

Figure 5. Kandinsky's Composition n°8.

We do not argue for a conflating of art with science. Indeed, calling aspects of science "art" can often be traced to feeble attempts at explaining poorly understood phenomena. The nexus between art and science perhaps can be best understood via the term often used in conjunction with the genesis of art, namely, "imagination." This can be appreciated in the often purported statement by Einstein that, "imagination is more important than knowledge." As the rest of the quotation continues, this is due to the expansion of the world view which imagination intimates as opposed to knowledge which is circumscribed. Yet Einstein himself was often required to change his mind when some of his predictions were not verified. This was the case when it was demonstrated that the universe was flying apart as opposed to being in a steady state.

Yet the differences (and similarities) between art and science continue to be viewed as being approaches to "knowing" of a unique sort. The Einstein experience would suggest that art (and imagination) are necessary for the scientific endeavour in terms of opening possibilities; nonetheless, it is science that must actually verify the possibilities. Thus, art and science both deal with knowledge of some sort processed through the human brain. And insofar as they share this common bedrock, they are both prone to the same maladies.

In the 19th century, it was fashionable among Marxist-materialist philosophers to emphasize the tendentious utility of art; i.e., art required a social purpose—often to support the prevailing ideology. And certainly science came under the same rubric. Marxism has had a diminished influence, but the tendency toward utility is still present. Implicit even in many scientific models is the idea that there is a teleology, either proximal or on a more grand scale. As soon as science makes such implicit assumptions, it is questionable if it is any longer science.

And more recently, "string theory" has engendered an emblematic discourse over this precise issue. String theory (via the mathematics of multidimensional spaces) has suggested that the universe is really composed of many dimensions. Yet the criticism is that such possibilities are not provable by experimental evidence. Thus, this is not science, but borders on the area of the mystical. Whether this is true will depend upon further explication, yet the very fact of this interchange between the "art" and "science" of knowledge points to the progressive and evolutionary nature of their mutual interaction.

REFERENCE

[1] The effect refers to the "morphing" to fit an image of self. See Yee N, Bailenson JN (2007, in press). The Proteus Effect: Self transformations in virtual reality. *Human Communication Research.* Another interesting paper explaining this point is Ioannidis JP (2005) Why most published research findings are false. *PloS Medicine* 2 (8): e124

Chapter 9

PROBABILITY AND DETERMINISM

God does not play dice with the universe.

Albert Einstein

Albert, stop telling God what to do.

Niels Bohr (Figure 1)

Not only does God play dice with the universe, but sometimes he throws the dice where they can't be seen

Stephen Hawking

Figure 1. Niels Bohr.

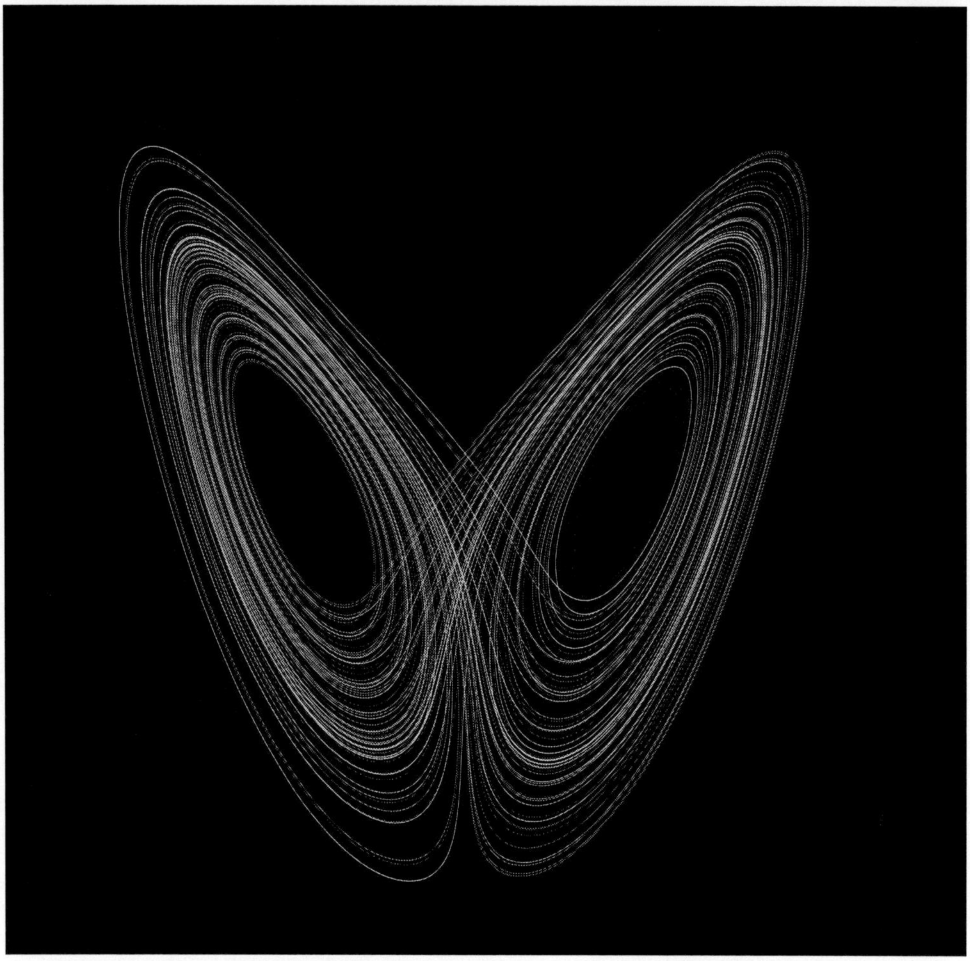

Figure 2. Lorenz attractor of chaos fame.

An underlying theme of the previous chapters has been the recognition that deterministic approaches to science often fail in the face of real data. This real data, then, is often appreciated through statistical approaches. This is especially true in dealing with biological and social sciences. Although determinism has its antecedents with Greek philosophy, the more proximal (and mathematically related) beginnings may start with Laplace's famous statement:

"We ought to regard the present state of the universe as the effect of its antecedent state and as the cause of the state that is to follow. An intelligence knowing all the forces acting in nature at a given instant, as well as the momentary positions of all things in the universe, would be able to comprehend in one single formula the motions of the largest bodies as well as the lightest atoms in the world, provided that its intellect were sufficiently powerful to subject all data to analysis; to it nothing would be uncertain, the future as well as the past would be present to its eyes. The perfection that the human mind has been able to give to astronomy affords but a feeble outline of such an intelligence. (Laplace 1820)"

The implication here is that there is a divine force knowing everything, with the human mind "feebly" outlining the intelligence. This obviously introduces a theological dimension.

Irrespective of whether one believes in such a force, LaPlace himself points out that human intelligence is feeble in this regard. Thus true prediction is impossible since no one possesses such an intelligence.

Nevertheless, as often happens, a statement that could suggest our impossibility to pursue the dream of "perfect control" was advocated for claiming the implicit possibility (even if in the long run) to get such a total knowledge. Moreover, every piece of scientific "progress" was more or less overtly described as a further step in the "total control" direction.

The pervasive charm of this dream of power which the ancient Greeks called ὕβρις (hubris, violence coming from an exaggerated self confidence about one's own possibilities), implying this feeling is at the basis of big catastrophes, contaminated virtually all fields of science. The horrors coming from this attitude are in front of us, just to name a few: colonialism, racism, eugenetics, ecological catastrophes, global climate change. The strength of this dream has been so big that it made many scientists forget that the major achievements of 20th century science gave exactly the opposite message, i.e., the intrinsic impossibility of such a total control: Poincare's work in dynamics, Gödel's theorem in mathematics, the development of quantum physics, from different vantage points demonstrated the intrinsic impossibility of such a dream (nightmare).

Even nowadays, there are resurgences of the deterministic obsession: the so-called "chaos theory" is a deterministic theory in which the probabilistic character is simply confined to our lack of knowledge (Figure 2). The lack of predictability simply stems from the fact we do not have "precise" knowledge of initial conditions. If we did possess infinite precision of these conditions, we could predict their motion.

The dream of perfect control was sustained by the successes of the physical sciences at least until the beginning of the 20th century. Until this time, scientists were able to develop differential equations which were able to predict various dynamics accurately. What was not appreciated was that this accuracy was specious. To a large extent, it was dependent upon scales of observation. Einstein's theory of relativity made clear that Newton's laws of motion were not, strictly speaking, accurate, but acceptable relative to the scale of observation. Indeed, when physical sciences became focused on the microscopic scales of observation (e.g., gas laws, quantum mechanics), a statistical appreciation was required. If anything, a series of statistical observations demonstrated mean values which might converge to a limit approaching a kind of determinism.

Whereas the physical sciences can be relatively easily be "isolated" for study, as we have pointed out, biological and social systems have great difficulty. The great confusion among these different spheres often redounds to the mathematics which is called to do the heavy work of modeling and prediction.

It has been pointed out that mathematics is study of mental objects with reproducible results; whereas the reproducible objects of the world constitute science. The great success of mathematics and science has been the frequent coincidence between the physical and mental worlds. This has by and large been based on the logic of mathematics, which is based on specific mental objects. This, historically, often appeared to be coincident. A change in the scale of observation changed this.

If decisions about physical objects (events) are no longer specified, but must be chosen from a probability developed from previous experience, then this is statistics and probability (dependent on physical objects)! The traditional syllogistic logic is no longer operative—instead, "random variables" need to be introduced. (There has been a great debate about the nature of random variables and measure theory. To discuss this would take away from the central point of this discussion). Thus it could be argued that statistics are the more natural approach for developing scientific models, since it is dependent on physical objects and not on idealized mental ones.

The discontinuity between mental and physical objects has led to many cures for the pathology, i.e., for increasing evidence of the failure to develop predictive models for real systems. This has included the inclusion of "noise terms" in equations modeling phenomena. Yet the "noise" itself is conceived as an ideal term. Real noise can have many different sources and of varying kinds. More importantly, real dynamics are rarely stationary, and the interplay between noise and "determinism" can change significantly. Furthermore, there is the very fundamental question regarding the source and function of real noise.

Insofar as science looks for patterns, and as has been pointed out, the human mind often creates patterns where there may be none, models must be constantly tested. In fact, a major problem with the idea of noise is that there has been a historical tendency to label things as noise which do not fit in with the conceived pattern. It is a sort of malevolence against nature that "refuses to collaborate" with the domain imposed by the human rational mind. Research over the last two decades, however, has pointed to the constructive role of noise in the area of "stochastic resonance," whereby a weak signal can be amplified by noise. This opens a complete reversal of attitude with respect to the world "outside" with scientists asking for a "collaboration" with the myriad of things which escape our explanation (noise), instead of considering them as an unwanted disturbance to their enterprise of perfect dominance. In our opinion this is a much more realistic (and thus in some sense rational) attitude than the deterministic obsession, blind to the failures coming from the application to reality.

An alternate conception of the universe is that it is composed of islands of patterns in a sea of noise, or that it is a system of nondeterminism and noise complicated by causality. With such a view, the "natural" measure would be probability and statistics. This view may be most important for the current concern regarding "complex systems" [1].

After decades of reductionism, scientists in many fields have turned their attention to see how the whole is greater than the "sum of its parts." Although the idea of complexity and emergence has found some favor in biological disciplines to describe the myriad processes found in organisms, many questions remain. The explosion of data concerning molecular biology has forced many fatigued scientists to ask if there are any over-all guiding principles, or "laws" akin to those sought by physicists and chemists. Implicit in this query is the additional question of how detailed must a model be for it to accurately describe a biological phenomenon. Clearly, some guiding principles could allow for reduced models which could be managed more easily. This is especially the case where mathematical models are the case. Currently, many mathematical models are high dimensional requiring considerable computational time. Yet the results of such models are often questioned for the obvious reason that even very small errors can quickly be magnified. There is also the very real "noise" problem: how does one model the naturally occurring noise found in biological systems? Different kinds of noise at different length scales makes the problem extremely important [2].

One can readily appreciate why many have invoked the "science of complexity" in this arena. If complexity exhibits a universal paradigm, and it is appropriate for the biological arena, then perhaps there may be a way of identifying guiding principles to reduce the dimensionalities of these biological problems. To this end, many have looked for signatures of this complexity, such as scale-free, network dynamics. Scale-free networks have been claimed for metabolic processes, organ systems, and regulatory structures. Very often these networks are determined in the context of "power laws," whereby some feature of the observed phenomenon falls off as an exponential power law.

The difficulty with such observations is that they are simply an observation of a kind of distribution, in the same way that it can be said that some variable exhibits a normal distribution, or a Poisson distribution. It says nothing in particular about the underlying mechanisms of the distribution. Indeed, such power laws are found in many phenomena and with a myriad of different mechanisms responsible for them—they are not necessarily unique.

MATHEMATICS AND PERCEPTIONS

Implicit in such observations is the belief that the mathematics reveals some underlying universal properties not clearly seen in the empirical observations. This is not surprising given that the foundation of classical mathematics is Platonist (Figure 3). Universals and laws are quite at home in Plato's cave. There is, however, a new challenge to this Platonistic foundation to mathematics, and it comes from one of the objects biologists wish to understand; namely, the human brain.

For a long time, it has been assumed that logic (or a logic) is independent of a thinker: "correct" reason is not dependent upon the peculiarities of a human brain--it transcends it and can be verified by any other correctly thinking human brain. However, in recent decades, cognitive neuroscientists have challenged this indirectly. Many studies have shown that the human brain loves patterns. And as David Ruelle has pointed out, our mathematics comes out of physical reality, filtered by the brain. In other words, we do the kind of mathematics we do because we "see" the mathematics in our reality—whatever this might be.

Figure 3. Plato.

This is an important consideration relative to the quest for biological universals, because it parallels the experience of physicists. Perhaps Heisenberg became the first widely known person to express it in his Uncertainty Principle, but the point is that an "observer" changes the thing observed by virtue of the observation. This is historically a remarkable statement which has often been glossed over by many scientists, although there have been a few who have attempted to comment upon it—mainly in the context of quantum physics. This observation puts into question the entire notion of "objectivity." And certainly, it has been the progenitor of many so-called "weird" effects of quantum science.

Although the biological experience is not quite the same as the physicist's, it points out, to a certain extent, that the scientist as observer, "filters" perceptions through the activity of the brain. Can this be the reason why so many scientists still pursue mechanistic mathematical models even when quantum physics has placed serious questions on such an endeavor? As is well known, strictly speaking, Newton's Laws are incorrect (as compared to quantum physics), but the errors are minimal and can thus be used. But does this gloss allow for a continued myopic view of various phenomena—especially in the life sciences. Consider that the Ptolemaic view of the Universe held sway because in terms of its predictions, the errors were few. The fact that biological systems must be continually adaptable, and in a sense "contingent" would tend to mitigate the universalist view. And what about the multiple causes of "noise" in the biological world. Often, noise is lumped into one big "black box" of unknowns and sequestered in differential stochastic equations.

Yet the biologist is confronted with the incontrovertible fact that higher biological systems coordinate many millions of processes simultaneously in a sea of noise to effect what are obviously deterministic processes. Thus neuronal synapses have a degree of randomness associated with them, as do ion channels, etc., yet they are capable of developing and coordinating "messages" that can ultimately result in such things as coordinated movement and thought. In biological systems, for instance in synaptic transmission, there is a random element, but this does not preclude a deterministic behavior at a higher level. In fact, while random behavior may sometimes be useful, this is usually not the case, and it seems that biological systems try to work deterministically even when starting with somewhat random processes. Where is Platonism in this picture? It may be just an idealization that is sometimes appropriate, and sometimes not, and is often accepted more for psychological than for scientific reasons.

SIMPLICITY

While complexity has been put forward as an important paradigm for biological systems, there is still "Occam's razor." And indeed, scientists of different fields are wont to quote this dictum of parsimony when deciding on multiple potential models. It was not however until the 1940s and 50s that there was significant attention paid to the idea of complexity. To be sure, if simplicity has been such an important guiding principle, how has its complement suddenly come to the fore? Clearly, simplicity has not been abandoned, inasmuch as such luminaries as Einstein and Hawking have invoked it. But where does complexity fit in?

Perhaps an analogy with medieval tapestries may clarify the issue: simple things are not necessarily more intelligible than complex ones. A tapestry is composed of numerous threads of different colors, which may be continuous throughout, but which appear at various points to trace out a figure or letter, etc (Figure 4). Thus apparent intricacy is really specious, and from single threads emerges a unique story. And certainly emergent properties are often cited as being a hallmark of complex systems: traits not seen in the individual parts are demonstrated only in the complete system. Yet what could be simpler than the periodic table of elements: proton number accounts for a periodicity and electron shell completeness to account for a very simple ordering of elements. From this basic arrangement emerge the elements' visibly different characteristics. It is still not clear how all this allows for such astonishingly different elemental attributes and perceptions. These attributes would appear to be "emergent" properties.

If the research on cognitive neurophysiology is to be believed, it may serve as a cautionary tale for a whole-hearted acceptance of bio-complexity. At the very least, the issue is far from decided in that a universally accepted definition of complexity is not at hand. What is known is that it is a big problem based on massive amounts of data. Still, it should be recalled that just because something is big, does not necessarily mean that it is complex; nor does it mean that just because something is small does not mean it is not complex.

Herbert Simon discovered that when confronted with massive amounts of data, the human mind was inclined to pull back to a smaller subset of data which was well appreciated to make decisions. A similar tendency may be operative with the suddenness of massive amounts of molecular data. There may be a psychological comfort in looking for "basic principles."

Figure 4. Bayeaux Tapestry.

Still there may yet be another approach. Given that there is at least some determinism, and some "noise," an approach which favors "piece-wise determinism" may provide a way which recognizes the essential "building block" nature of many processes, while at the same time recognizing that building blocks are not always easy to assemble.

In the first chapter we pointed out that if the Lipschitz conditions for differential equations are relaxed, this allows for a simpler way to construct processes. At the same time it can more easily model the massive asynchronous parallelism found in biological processes [3]. Indeed, when it comes to modeling neural nets with their asynchronous, and massively parallel "computation," non-Lipschitz equations may be more appropriate. Traditional "neural net" models of cognition have some difficulties, including difficulties dealing with noise, parallelism, etc. A "piece-wise deterministic," or "terminal" model of neurodynamics can exhibit a stochastic attractor as a universal tool for generalization (Figure 5).

Indeed, in contradistinction to chaotic attractors of deterministic dynamics, stochastic attractors can provide any arbitrarily prescribed probability distributions by an appropriate choice of (fully deterministic!) synaptic weights. Certainly such a dynamical system is capable of discrimination between different stochastic patterns, and therefore, it performs pattern recognition on the level of classes.

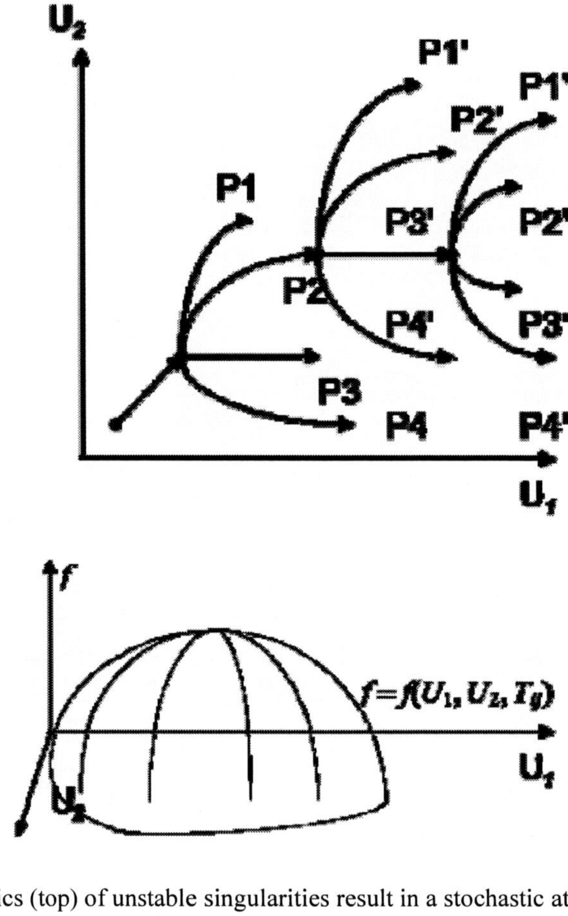

Figure 5. Combinatorics (top) of unstable singularities result in a stochastic attractor (bottom).

Another example is the interest that has been paid to metabolic scale free networks. The architecture of these metabolic nets is in many ways similar to neural nets. Indeed, several studies based on real data point to lognormal distributions and power laws. Thus there is a good reason to consider the possibility of underlying multiplicative processes. Furthermore it should be stressed that the benefit of piece-wise modeling is that such dynamics are much more forgiving of "errors."

What is notable about piece-wise determinism is that it acknowledges that a given corpus can constitute a very large amount of inter-related data, while at the same time posing the linkages in a way which does not tie the deterministic pieces to the past. A further benefit is that "control" of the dynamics is exquisite at the singularities, while being relatively impervious to the effects of noise once a trajectory is chosen. In some sense, one can view this model as biological "optimization." The stochastics of one level evolve into a progressively "deterministic" process at a higher level as can be seen through lognormal statistical distributions. Thus a very large problem can be reduced to one of combinatorics and probabilities. This last point needs to be emphasized since there is a persistent tendency to view many of the implied dynamics through the legacy of Laplace. This is to say that there is a tendency to model these dynamics with traditional differential equations. As has been suggested, however, this need not be so. Piece-wise deterministic dynamics can produce similar results, but are relatively simpler, and perhaps heuristically more satisfying, given the underlying stochastics [4]. A related approach is that of "swarm theory," which has gained many adherents in the optimization community. Essentially, by mimicking rules that govern, e.g., ant colonies, it has been found that many so-called "complex behaviors" are based on very simple rules [5]. Piece-wise dynamics differs somewhat by emphasizing the probabilistic nature of the singularities.

One topic which we have not discussed is the fact that most real models do not specifically consider the fact that most real systems are "dissipative," i.e., they are "open" systems, losing energy, or exchanging energy from the outside. Obviously, this is especially true for biological systems. "Optimization" in this context can be problematic—where is energy lost or gained? Modeling from this perspective almost invites over-determination, and may be best dealt by statistical means and the related probabilities. Piece-wise dynamics may be uniquely suited for such an approach [6].

The global challenges mankind now faces in terms of climate change with the previsions of planetary catastrophes that no more allow for continuing the trend of an "artificial world" largely independent of the natural dynamics and compatibilities makes this change of attitude with the consequent acquiring of a much more humble view toward natural phenomena urgent. We hope this little book could be an infinitesimal (but nevertheless different from zero) step toward this direction.

REFERENCES

[1] For a further discussion, see Zbilut JP, Hubler A, Webber Jr CL (1996). Physiological singularities modeled by nondeterministic equations of motion and the effect of noise. In M Millonas (ed), *Fluctuations and Order: The New Synthesis*. Springer Verlag, NY, pp 397-417.

[2] EF Keller engendered some interesting commentaries in his “A Clash of Two Cultures” *Nature* 445: 603 (2007).

[3] Zak M, Zbilut JP, Meyers RE (1997). *From Instability to Intelligence: Complexity and Predictability in Nonlinear Dynamics, (Lecture Notes in Physics: New Series m 49)*. Springer Verlag, Berlin Heidelberg New York. Also Zbilut JP (1984) *Unstable Singularities and Randomness*. Elsevier, Amsterdam.

[4] That such a scenario is plausible has been demonstrated by JA Bradford, KA Dill (2007). Stochastic innovation as a mechanism by which catalysts might self-assemble into chemical reaction networks. *Proceedings of the National Academy of Sciences (USA),* 104: 10098-10103.

[5] Miller P. (2007). Swarm theory. *National Geographic*, July (2007), pp. 126-147.

[6] Optimization is historically related to the idea of the “principle of least action.” Many subtleties are involved in this topic, and not appreciated by many contemporary scientists. An excellent introduction is found in Ekeland I (2006). *The Best of All Possible Worlds*. University of Chicago Press, Chicago. His discussion is primarily related to the physical sciences, and only marginally discusses biological examples.

Conclusion: Bringing it All Back Home: An Inconclusive Conclusion

Complexity is simple—like drinking water through a fire hydrant.

John F. Hibbeln

As with any cafeteria debate, even this must come to an end. When people leave the table and say goodbye to each other, we are left with the doubt if the correct message was clear.

The previous statement is perhaps a bit conceited by implying the message is something that is definitely there, generated by the mind of the writers, but that the receiver may or may not correctly decode it [engineers have very effective statistical methods called ROCs (Receiver Operating Curves)] to give a mathematical score to the efficiency of the correct receive of a signal). But our "message" does not imply that the reader acts as a mechanical device (like in ROCs). The message, without an active and critical reader does not exist at all, so any reader will leave the cafeteria with his own idea about the things he listened to.

We are sorry for the repetitions of concepts. [We have an unfortunate fondness for the maxim, "repetitio est mater studiorum" (repetition is the mother of study)] The authors are bit maniacal with two or three "basic statements" their friends know very well and are benignly sarcastic about this. The "take home message" is thus different for different homes, but we hope any reader that had the patience to read the entire book got the impression of how science can profitably advance by being considered a form of subtly simple "artistic craft" instead of an arrogant "unique view of the world."

PICTURE CREDITS

CHAPTER 1

Michael Jordan by Joshua Massel. Some rights reserved under *http://creativecommons.org /licenses/by-sa/2.0/*

CHAPTER 2

Figure 6. David Ruelle. Photo by Vojkan Jaksic (Math Dept, McGill, Montreal).

Figure 8. Kandinsky's Autumn in Bavaria. Created by Nicolas Pioch, with permission through Creative Commons Attribution-ShareAlike 2.5 License.

CHAPTER 5

Drawing by Julian Beever. *http://users.skynet.be/J.Beever/postmod.htm* By permission.

CHAPTER 6

Figure 1. N. Wiener. Image courtesy of the Research Laboratory of Electronics at MIT.

INDEX

E

F

G

H

I

J

K

L

M

N

O

P

Q

R

S

T

U

V

W

Y

Z